PROTECTING PARTICIPANTS AND FACILITATING SOCIAL AND BEHAVIORAL SCIENCES RESEARCH

Panel on Institutional Review Boards, Surveys, and Social Science Research

Constance F. Citro, Daniel R. Ilgen, and Cora P. Marrett, *Editors*

C

Board on Beh[...]   [...]ciences

Division on Beh[...]   [...]a Social Sciences and Education

NATIONAL RESEARCH COUNCIL
OF THE NATIONAL ACADEMIES

THE NATIONAL ACADEMIES PRESS
Washington, D.C.
www.nap.edu

THE NATIONAL ACADEMIES PRESS   500 Fifth Street, NW   Washington, DC 20001

NOTICE: The project that is the subject of this report was approved by the Governing Board of the National Research Council, whose members are drawn from the councils of the National Academy of Sciences, the National Academy of Engineering, and the Institute of Medicine. The members of the committee responsible for the report were chosen for their special competences and with regard for appropriate balance.

The project that is the subject of this report was supported by contract no. SBR-9709489 between the National Academy of Sciences and the National Science Foundation. Any opinions, findings, conclusions, or recommendations expressed in this publication are those of the author(s) and do not necessarily reflect the views of the organizations or agencies that provided support for the project.

International Standard Book Number 0-309-08852-6 (book)
International Standard Book Number 0-309-51136-4 (PDF)
Library of Congress Control Number: 2003106396

Additional copies of this report are available from the National Academies Press, 500 Fifth Street, NW, Washington, D.C. 20001; (202) 334-3096; Internet, http://www.nap.edu

Suggested citation: National Research Council (2003). *Protecting Participants and Facilitating Social and Behavioral Sciences Research.* Panel on Institutional Review Boards, Surveys, and Social Science Research. Constance F. Citro, Daniel R. Ilgen, and Cora B. Marrett, eds. Committee on National Statistics and Board on Behavioral, Cognitive, and Sensory Sciences. Washington, DC: The National Academies Press.

THE NATIONAL ACADEMIES
Advisers to the Nation on Science, Engineering, and Medicine

The **National Academy of Sciences** is a private, nonprofit, self-perpetuating society of distinguished scholars engaged in scientific and engineering research, dedicated to the furtherance of science and technology and to their use for the general welfare. Upon the authority of the charter granted to it by the Congress in 1863, the Academy has a mandate that requires it to advise the federal government on scientific and technical matters. Dr. Bruce M. Alberts is president of the National Academy of Sciences.

The **National Academy of Engineering** was established in 1964, under the charter of the National Academy of Sciences, as a parallel organization of outstanding engineers. It is autonomous in its administration and in the selection of its members, sharing with the National Academy of Sciences the responsibility for advising the federal government. The National Academy of Engineering also sponsors engineering programs aimed at meeting national needs, encourages education and research, and recognizes the superior achievements of engineers. Dr. Wm. A. Wulf is president of the National Academy of Engineering.

The **Institute of Medicine** was established in 1970 by the National Academy of Sciences to secure the services of eminent members of appropriate professions in the examination of policy matters pertaining to the health of the public. The Institute acts under the responsibility given to the National Academy of Sciences by its congressional charter to be an adviser to the federal government and, upon its own initiative, to identify issues of medical care, research, and education. Dr. Harvey V. Fineberg is president of the Institute of Medicine.

The **National Research Council** was organized by the National Academy of Sciences in 1916 to associate the broad community of science and technology with the Academy's purposes of furthering knowledge and advising the federal government. Functioning in accordance with general policies determined by the Academy, the Council has become the principal operating agency of both the National Academy of Sciences and the National Academy of Engineering in providing services to the government, the public, and the scientific and engineering communities. The Council is administered jointly by both Academies and the Institute of Medicine. Dr. Bruce M. Alberts and Dr. Wm. A. Wulf are chair and vice chair, respectively, of the National Research Council.

www.national-academies.org

v

Acknowledgments

The Panel on Institutional Review Boards, Surveys, and Social Science Research thanks the many people who contributed their time and expertise to the preparation of this report.

We are grateful to everyone who attended the panel's first meeting and provided perspectives on issues of human research participant protection in the social, behavioral, and economic sciences (SBES). We acknowledge the wealth of information that we obtained from websites of private and public organizations and from previous surveys of the review process for research with human participants (see the appendices). We also acknowledge the very useful paper by George Duncan, of Carnegie Mellon University, on confidentiality and data access issues for institutional review boards (IRBs), which is reproduced as Appendix E. We thank the staff of the National Research Council for their helpful advice and input, including Andrew White, director of the Committee on National Statistics; Christine Hartel, director of the Board on Behavioral, Cognitive, and Sensory Sciences; and Laura Lyman Rodriguez and Jessica Aungst, staff to the Institute of Medicine Committee on Assessing the System for Protecting Human Research Participants. That committee produced the very useful report, *Responsible Research*, which provides an invaluable perspective on the work of IRBs, primarily in the biomedical fields. Eugenia Grohman, director of the reports office of the Division of Behavioral and Social Sciences and Education, made important contributions to our report through her fine technical editing.

Our panel was assisted by a very able staff. We are grateful to Virginia A. deWolf, now at the U.S. Bureau of Transportation Statistics, who served as the panel's first study director. She did a masterful job of organizing the panel's meetings, reaching out to other groups that are active in human research participant protection issues, and assembling a wealth of background materials to inform the panel's deliberations. Jamie Casey conducted the panel's review of websites of 47 major research institutions to determine their guidance and policies for review of research protocols involving human participants, tracked down often obscure materials for the panel's use, and assisted the panel at its meetings. Tanya Lee made excellent arrangements for panel meetings. Daniel Cork contributed his outstanding typographic skills to the preparation of the report for printing.

The panel is especially grateful to Constance Citro, who served as the panel's study director beginning in May 2002. She insisted that our work reflect the highest standards of evidence and worked unfailingly to uncover sources for that evidence. We draw attention in particular to the synthesis she developed on the evolution of federal guidelines for the protection of human participants in research, which is recounted in Chapter 3. From disparate sources, she developed a coherent and original account of that process. More broadly, with extraordinary diligence, she managed the completion of the panel's work.

I want to extend special thanks to Daniel Ilgen, who served as vice chair of the panel. He assumed the role despite a lengthy list of other commitments. He listened to our deliberations and crafted arguments noteworthy for their clarity. He worked tirelessly with Connie to ensure that our efforts warranted the imprimatur of the National Research Council.

All of the panel members made important contributions of their time and expertise, not only bringing to bear examples and perspectives from their own specialties, but also engaging in intensive dialogue to reach consensus on key issues for participant protection in SBES research. It was an honor to serve with them.

The panel also benefited from our two liaison members. William Yost, Loyola University, Chicago, liaison from the Board on Behavioral, Cognitive, and Sensory Sciences to our panel, attended all our meetings and provided a very useful perspective to the panel's deliberations. Roderick J.A. Little, University of Michigan, attended our early meetings as liaison from both the IOM committee and the Committee on National Statistics.

This report has been reviewed in draft form by individuals chosen for their diverse perspectives and technical expertise, in accordance with procedures approved by the National Research Council's Report Review Committee. The purpose of this independent review is to provide candid and critical comments that will assist the institution in making its published report as sound as possible and to ensure that the report meets institutional standards for objectivity, evidence, and responsiveness to the study charge. The review comments and draft manuscript remain confidential to protect the integrity of the deliberative process.

We wish to thank the following individuals for their review of this report: Evan G. DeRenzo, Center for Ethics, Washington Hospital Center, Washington, DC; Lowell W. Gerson, Office of Addiction Medicine, Northeastern Ohio Universities College of Medicine, Rootstown, OH;

Jeff Kahn, Center for Bioethics, University of Minnesota; Richard A. Kulka, Research Triangle Institute, Research Triangle Park, NC; Roderick J.A. Little, Department of Biostatistics, University of Michigan; Richard E. Nisbett, Culture and Cognition Program and Department of Psychology, University of Michigan; Lee N. Robins, Department of Psychiatry, Washington University School of Medicine, St. Louis, MO; and Joan E. Sieber, Department of Psychology, California State University, Hayward.

Although the reviewers listed above have provided many constructive comments and suggestions, they were not asked to endorse the conclusions or recommendations nor did they see the final draft of the report before its release. The review of this report was overseen by Henry W. Riecken, Behavioral Sciences, University of Pennsylvania, and Mary Jane Osborn, Department of Microbiology, University of Connecticut Health Center. Appointed by the National Research Council, they were responsible for making certain that an independent examination of this report was carried out in accordance with institutional procedures and that all review comments were carefully considered. Responsibility for the final content of this report rests entirely with the authoring committee and the institution.

Cora B. Marrett, *Chair*
Panel on Institutional Review Boards,
Surveys, and Social Science Research

Contents

List of Figures

List of Boxes

Executive Summary

THE U.S. SYSTEM for protecting people who volunteer to participate in research is widely perceived to need improvement. A major concern is that the linchpins of the protection system—institutional review boards (IRBs)—are overloaded and underfunded and so may not be able to adequately protect participants from harm in high-risk research, such as clinical trials of experimental drugs.

Three other concerns—often voiced about research in the social, behavioral, and economic sciences (SBES), but generally applicable to human participant protection—are important. The first is that the review process too often focuses on documenting consent to participate in research so as to satisfy the letter of federal requirements, when IRBs and researchers instead need to focus on developing the most effective processes for helping individuals reach an informed, voluntary decision about participation. The second concern is that IRBs, researchers, and the entire human participant protection system may pay too little attention to the challenge of countering increasing threats to the confidentiality of research data because of technological and other changes, such as the ability to readily access and link large databases through the Internet. The third concern is that the review process may delay research or impair the integrity of research designs, without necessarily improving participant protection, because the type of review is not commensurate with risk—for example, full board review for minimal-risk research that uses such methods as surveys, structured interviews, participant observation, laboratory experiments, and analyses of existing data.

PANEL CHARGE AND SCOPE

The Panel on Institutional Review Boards, Surveys, and Social Science Research was established by the Committee on National Statistics and the Board on Behavioral, Cognitive, and Sensory Sciences, both standing committees of the National Academies' National Research Council. The panel was charged to examine the structure, function, and performance of the IRB system as it relates to SBES research and to recommend research and practice to improve the system.

1

Our panel's work complements that of the Institute of Medicine's Committee on Assessing the System for Protecting Human Research Participants, which issued its final report, *Responsible Research*, in 2002. That report addresses primarily the problems of high-risk research. We commend that report, which stresses that participant protection in the United States is a dynamic system of many actors. The report makes useful recommendations to virtually all actors, including the Congress, the Office for Human Research Protections (OHRP) in the U.S. Department of Health and Human Services, agencies that support research and collect data for research use, high officials of research institutions, IRBs, researchers, individual participants, and many interested associations and other organizations.

Although addressed primarily to SBES research, our findings and recommendations have broader application given that the boundaries between research domains are not and cannot be sharply drawn. We address three topics in depth: issues for obtaining informed, voluntary consent; issues for protecting data confidentiality; and review procedures for minimal-risk research. We consider more briefly system-level issues regarding the relationships and interactions among actors involved in participant protection. Throughout, we stress commitment to upholding the principles for ethical research articulated in the landmark 1979 Belmont Report: respect for persons (informed consent), beneficence (minimizing risks and maximizing benefits of research), and justice (selection of participants in ways that fairly distribute the burdens and benefits of research). Given scarce IRB resources, we believe that commitment to protection requires that review procedures be commensurate with risk. The Common Rule regulations ("Federal Policy for the Protection of Human Subjects") contain sufficient flexibility for this purpose: the challenge is how best to encourage IRBs to use the flexibility in the regulations appropriately for different types of research methods, topics, and study populations.

ENHANCING INFORMED CONSENT

Informed consent is a bedrock principle of ethical research with human participants. For more-than-minimal-risk research, a process that allows consent to be truly informed is critical; for minimal-risk research, such a process respects individual autonomy. Despite decades of research on consent issues, mostly in biomedical research and mostly involving written forms, there appears to have been little progress in devising more effective forms and procedures for achieving informed consent or in adapting consent procedures to the needs of special populations (e.g., language minorities).

Recommendation 4.1: Social, behavioral, and economic science researchers should conduct research on procedures for obtaining and documenting informed consent that will facilitate comprehension of benefits, harms, and risks of harm, confidentiality protection, and other key features of research protocols for different types of SBES research and populations studied.

Recommendation 4.2: The Office for Human Research Protections should develop detailed guidance for IRBs and researchers on appropriate consent procedures for different types of populations—including language minorities and such vulnerable groups as undocumented immigrants—studied in social, behavioral, and economic sciences research.

The issue of third-party consent has gained salience in recent years due to reports of studies in which third parties complained that their privacy was invaded by collection of sensitive data about them from others. Examples of research that should not require third-party consent, even though information about third parties is sought, are studies in which respondents are asked about their perceptions or attitudes regarding others, studies in which the third person asked about is completely anonymous (e.g., a respondent's first teacher), and studies that present no more than minimal risk for third parties.

Recommendation 4.3: The Office for Human Research Protections should develop detailed guidance for IRBs and researchers, including specific examples, on when it is and is not necessary to obtain consent from third parties about whom participants are asked to provide information.

The current preoccupation of the review process with the documentation of consent may shift attention from protecting participants to protecting the research institution. Requiring a signed consent form for all types of research may inhibit participation in minimal-risk research (e.g., mail surveys of the general adult population) by otherwise willing candidates. In some situations, requiring signed written consent may endanger participants when there is risk of serious harm from breaching confidentiality and the only link of participants to the project is the signed consent form. The Common Rule allows for waiver of written signed consent when appropriate for minimal-risk research; it also allows elements of informed consent (e.g., the purpose of a particular aspect of the research) to be omitted under certain circumstances.

Recommendation 4.4:   The Office for Human Research Protections should develop detailed guidance for IRBs and researchers—with clear examples for a variety of social, behavioral, and economic sciences research methods and study environments—on when it is appropriate to waive signed written consent.

Recommendation 4.5: The Office for Human Research Protections should develop detailed guidance for IRBs and researchers, including specific examples, on when it is acceptable to omit elements of informed consent in social, behavioral, and economic sciences research.

ENHANCING CONFIDENTIALITY PROTECTION

Breach of confidentiality, that is, the release of data that permit identifying an individual participant, is often the major source of potential harm to participants in SBES research. For example, a survey that poses no risk of physical injury and no more than minor psychological annoyance may yet obtain data that could adversely affect a respondent's employability, insurability, or other aspects of life if it became known. Even if no sensitive information is obtained, maintaining confidentiality is required to respect participants when they have been assured that their information will be protected.

The risk of inadvertent or advertent disclosure is increasing due to several factors: the growing number and variety of administrative records from public and private agencies that are readily available on the Internet and potentially linkable to research data; the growing number of rich, longitudinal data sets that require retention of contact information for respondents over long periods of time and that may be more readily linked to other data sources with sophisticated matching techniques; the increased emphasis by funding agencies on data sharing among researchers to permit replication and facilitate further research at low cost; and the increased use of Internet-based data collection technology that may be vulnerable to security breaches.

Recommendation 5.1: Because of increased risks of identification of individual research participants with new methods of data collection and dissemination, the human research participant protection system should continually seek to develop and implement state-of-the-art disclosure protection practices and methods. Toward this goal:

- researchers should explicitly describe procedures to protect the confidentiality of the data to be collected in protocols they submit to IRBs;

- IRBs should pay close attention to the adequacy of proposed procedures for protecting confidentiality;

- federal funding agencies should support research on techniques to protect the confidentiality of SBES data that are made available for research use; and

- the Office for Human Research Protections should regularly promulgate good practices in analyzing disclosure risks and limiting those risks.

Increased attention to confidentiality protection does not mean that IRB review is needed for every type of analysis. Anecdotal evidence suggests that many IRBs are reviewing research with publicly available microdata files, even though such research qualifies for exemption. Such review uses up scarce IRB and investigator resources yet is unlikely to afford greater protection to respondents than is already incorporated in the design and content of the file.

Recommendation 5.2: To facilitate secondary analysis of public-use microdata files, the Office for Human Research Protections, working with appropriate federal agencies and interagency groups, should establish a new confidentiality protection system for these data. The new system should build upon existing and new data archives and statistical agencies.

Recommendation 5.3: Participating archives in the new public-use microdata protection system should certify to researchers whether data sets obtained from such an archive are sufficiently protected against disclosure to be acceptable for secondary analysis. IRBs should exempt such secondary analysis from review on the basis of the certification provided.

EFFECTIVE REVIEW OF MINIMAL-RISK RESEARCH

The work of IRBs begins with four sequential decisions about research projects:

(1) whether the project constitutes "research" under the Common Rule;

(2) whether it involves "human participants;"

(3) whether it falls into one of the specified categories that are exempt from IRB review; and

(4) if it is not exempt, whether it is minimal risk and eligible for review by the chair or subcommittee (expedited review) rather than the full board.

In the current environment of heightened scrutiny of IRB operations because of serious harms (even death) to research volunteers, IRBs often opt for full board review of minimal-risk research, even when such review is not appropriate or necessary for protection of participants and detracts from the attention needed for more-than-minimal-risk research.

More detailed guidance on review of minimal-risk research can encourage IRBs to use the flexibility in the regulations in an appropriate way. It can also reduce the substantial variability among IRBs in the use of such procedures as expedited review and so facilitate multisite research and make it easier for researchers to carry projects from one institution to another without encountering very different IRB standards. Such guidance should include clear examples for a variety of methods and populations studied. For example, research with publicly available aggregate data (e.g., tallies of census data for cities) does not involve human subjects under the regulations, and research with publicly available microdata of individual records qualifies for exemption when the data are certified by the supplier agency to be protected against breach of confidentiality.

> **Recommendation 6.1:** To promote review appropriately tailored to risk, the Office for Human Research Protections should develop detailed guidance for IRBs and researchers (with clear examples for a variety of methods) on what kinds of social, behavioral, and economic sciences (SBES) research protocols qualify as "research" with "human subjects." OHRP should also develop detailed guidance, including examples, regarding SBES research that IRBs are strongly encouraged to exempt from review and research that IRBs are strongly encouraged to review with an expedited procedure.

> **Recommendation 6.2:** Institutional review boards should use efficient procedures to review minor changes to minimal-risk research protocols that arise during the period of authorization. When appropriate, IRBs should approve protocols that allow researchers flexibility in making specific design

decisions during the course of their research without the need to seek further review. (An example would be one of two forms of a question—both minimal risk—to be decided on the basis of a pretest.)

NEEDED INFORMATION

We found, as did the Institute of Medicine study, that there is little regularly available systematic information about the functioning of the U.S. human research participant protection system. Data on harms encountered by research participants and their economic and other costs are scant. Only a handful of major surveys, smaller surveys, and case studies have examined IRB operations and the consequences for participant protection and timely research.

> **Recommendation 6.3:** In order to build knowledge of research risks, OHRP and funding agencies should encourage researchers to build into their studies such steps as debriefing participants to learn about types, incidence, and magnitude of harm encountered in social, behavioral, and economic sciences research. Researchers should seek publication of their results.

> **Recommendation 6.4:** The Office for Human Research Protections should establish an ongoing system for collecting and publishing data that can help assess how effectively IRBs protect human research participants, how efficiently they review research, and how commensurate review is with risk.

> **Recommendation 6.5:** Federal research funding agencies, including the National Science Foundation and the National Institutes of Health, should fund in-depth studies to better understand the operations and effects of the IRB system and to develop useful indicators of IRB performance.

SYSTEM-LEVEL ISSUES

The U.S. system for human research participant protection involves many components and is dynamic, evolving as social and economic changes affect various system components and they in turn respond. We consider five system-level issues that need continued attention: (1) guidance and support for IRBs; (2) qualifications and performance standards for IRBs and researchers; (3) communication among IRBs and researchers; (4) organization of and among IRBs; and (5) the development of national policy for human research participant protection. In most instances, we endorse recommendations of other groups, such as the Institute of Medicine and the National Bioethics Advisory Commission. In two areas that are particularly important for SBES research we offer recommendations.

Recommendation 7.1: To improve IRB-researcher communication and facilitate the review process, IRBs should:

- clearly distinguish and justify changes to research designs that are required for human participant protection from suggested changes that are advisory; and

- develop ways to work cooperatively with investigators, such as providing opportunities for face-to-face meetings to discuss significant changes in research protocols that the IRB requires.

Recommendation 7.2: Any committee or commission that is established to provide advice to the federal government on human research participant protection policy should represent the full spectrum of disciplines that conduct research involving human participants. In particular, such a body should include members who represent the range of the social, behavioral, and economic sciences.

The benefits of involving the SBES community should include not only increased support for and understanding of human participant protection policies among SBES researchers, but also useful cross-fertilization of knowledge and practice between SBES and biomedical researchers and IRB members. Such cross-fertilization will help the protection system better shoulder the difficult tasks of facilitating informed consent, protecting confidentiality, estimating risk, and taking other steps to fully protect and respect the many millions of Americans who have volunteered to participate in research to advance knowledge.

— 1 —
Introduction

PROGRESS IN UNDERSTANDING people and society and in better-ing the human condition depends on people's willingness to participate in research. In turn, involving people as research participants carries ethical obligations to respect their autonomy, minimize their risks of harm, maximize their benefits, and treat them fairly.

The U.S. government instituted policies designed to protect human research participants in the 1960s. Those policies, which gained regulatory force beginning in 1974, have evolved over the past 40 years in response to the concerns of Congress, executive agencies, researchers, and the public. Often those efforts were energized by media reports of unethical, even life-threatening, research. Today, most federal agencies that fund or conduct research on humans have adopted the "Federal Policy for the Protection of Human Subjects," known as the Common Rule (see Box 1-1, at the end of the chapter). The Common Rule lays out a set of protections and related requirements applicable to all research on human participants that is conducted, funded, or overseen by federal agencies or conducted at institutions receiving federal funds that have agreed to these protections for all research at their sites.[1]

The Common Rule provides for the establishment of institutional review boards (IRBs) to review and monitor individual research projects with human participants. It charges IRBs to assess harms, risks, and benefits of proposed research and to protect participants by requiring investigators to follow appropriate informed consent procedures and other procedures. It distinguishes "minimal-risk" research (see Chapter 2), which may receive an expedited IRB review, from research that is subject to full review (see Box 1-2, at the end of the chapter).

Key actors in the U.S. human participant protection system, with legal obligations under the Common Rule, are federal agencies that sponsor research with human participants, the Office for Human Research Protections (OHRP) in the U.S. Department of Health and Human Services (DHHS), officials of institutions that conduct federally funded research, research investigators, and IRBs, of which there were

---

[1]The Common Rule does not apply to privately funded research conducted or sponsored by organizations that receive no federal funds or oversight.

9

an estimated 4,000 in 1998 (Gunsalus, 2001:fn6). Other actors in the system include federal agencies and other organizations that provide data for research use, scientific professional associations, advisory committees to federal agencies, Congress, advocacy groups for participants, and associations of research organizations.

THE ISSUES

At present, the participant protection system is widely perceived to need improvement. From the perspective of high-risk clinical research (e.g., trials of experimental drugs), a primary concern is that the linchpins of the system—IRBs—may not be able to provide a sufficient level of review to protect research participants from serious injury and even death.[2] Systemic problems of underfunding and work overload of IRBs are believed to be major contributors to this situation. Also, in this view, federal regulatory agencies, until recently, have been lax in their oversight of IRBs (see, e.g., Office of Inspector General, 1998b).

From the perspective of research in the social, behavioral, and economic sciences (SBES), the generic concerns with the participant protection system are the same as in all disciplines—namely, that the system protect research participants as fully as possible while not placing more burdens on the conduct of useful research than are necessary to ensure that the research is ethical. In addition, three specific concerns are paramount (although not unique) to SBES research.

The first concern involves the practice of informed consent. Although there is wide consensus on the role of informed consent in protecting and respecting the rights of human research participants, in practice the IRB review system appears to pay too little attention to the process of helping individuals decide about participation and too much attention to documenting consent to formally satisfy federal requirements. Often, the result is the creation of a consent form with hard-to-understand "boilerplate" language that does not really enable prospective participants to appropriately assess the risks and benefits of participation (see, e.g., Sieber, Plattner, and Rubin, 2002).

A second concern is that new information storage and retrieval technologies, such as the Internet, are challenging traditional practices for protecting the confidentiality of research data. It is likely that those involved in the human participant protection system, including IRBs and investigators, are paying too little attention to the ways in

---

[2]A recent example was the death of a healthy young woman participant in a study about asthma medications. In response, OHRP suspended research involving human participants at Johns Hopkins University, regardless of risk (Keiger and De Pasquale, 2002).

which technological and other changes in the research environment are increasing the risk of disclosure of the identity of participants in research (see, e.g., Sweeney, 2001). Such disclosure is often the primary risk to participants in SBES research.

A third concern is that increasing scrutiny of and pressures on IRBs and the research institutions in which they exist are creating a bias toward overly protective review practices. This bias is likely to influence choices about informed consent practices. It is also likely to lead to choices by IRBs to subject research protocols to levels of review, such as full board review, that are not needed to protect participants in research that poses no more than minimal risk of harm. More stringent review than is necessary often delays research, sometimes results in inappropriate changes in research designs, and creates cynicism in the research community about the proper role of human participant protection (see, e.g., American Association of University Professors, 2001). Moreover, because IRB resources are limited, full review of minimal-risk research contributes to the burden on IRBs and limits their ability to devote sufficient attention to research that needs fuller scrutiny.

PANEL CHARGE AND SCOPE

In 2001 the Committee on National Statistics, in collaboration with the Board on Behavioral, Cognitive, and Sensory Sciences (both standing committees of the National Academies' National Research Council), established our Panel on Institutional Review Boards, Surveys, and Social Science Research. Our panel's charge was to examine the structure, function, and performance of the IRB system as it relates to SBES research and to recommend research and practice to improve the system.

Our panel's work was intended to complement the work of the Committee on Assessing the System for Protecting Human Research Participants of the Institute of Medicine (IOM), which issued its final report, *Responsible Research*, in October 2002. When the IOM committee began its study in 2000, it was intended to focus its attention primarily on biomedical research, particularly such high-risk research as clinical trials of experimental medical treatments, which have generated most of the public concern. Therefore, our panel was to pay particular attention to SBES research using such methods as surveys, structured interviews, participant observation, laboratory and field experiments, and analyses of existing data.

We commend the IOM report, which stresses that research participant protection in the United States is a dynamic system of many actors. The report contains recommendations for almost every actor in the system, including the Congress, OHRP, high officials of research institutions, IRBs, researchers, and individual participants themselves.[3] The IOM recommendations are designed to improve the performance of the participant protection system. Many of them, such as a recommendation for better data on research injuries, are also designed to bolster public trust that the system minimizes the risks of harm to participants as far as is humanly possible and also enables valuable, ethically responsible research to proceed. No research can be totally risk free, but the public deserves to know that the system for protecting research volunteers is operating well; further, that the system is capable of adapting expeditiously to changes in the research environment that call for changes in the protection system.

The IOM report includes as Appendix B a letter, dated July 1, 2002, that our panel sent to the IOM committee to provide input to the committee's deliberations from an SBES perspective. The letter provided our panel's initial recommendations on four topics: requirements for informed consent, particularly for advance written consent; protection of confidentiality of information obtained from participants; procedures for determining what research should be exempt from IRB review or should receive expedited review; and system-level issues, such as training of researchers and accreditation.

This, our final report, elaborates on the issues raised in our letter and provides more extensive background information and supporting material. In it, we have adopted the systems perspective of the IOM committee's report, which recognizes that appropriate interactions among all of the components of the participant protection system are necessary for the system to operate responsibly and effectively. Because of the time and resources available for the panel's work, we do not address every aspect of the protection system. We address some issues and perspectives briefly (e.g., education and training for IRBs and researchers) and three topics in depth:

- We consider informed consent in terms of the obligation of IRBs to focus more on the consent process and less on the consent form, the obligation of SBES researchers to conduct research on effective consent processes and documentation, and the obligation of OHRP to provide guidance that will enable IRBs and researchers to make participant protection paramount in considering consent procedures for specific research protocols.

---

[3]Except in citing regulatory language, we have followed the IOM report language of human research "participants" instead of the more commonly used term, "subjects."

- We consider confidentiality protection in terms of the need for IRBs, investigators, and OHRP to recognize the increasing risks of disclosure in today's computing environment with the ready availability of rich data files on the Internet and sophisticated matching software. We emphasize the role that federal statistical agencies and data archives can play in providing access to data while minimizing the risk of disclosure.

- We consider review procedures for minimal-risk research in terms of the obligation of researchers to build a body of evidence about perceived and actual harms in SBES research, the obligation of IRBs to review research at a level commensurate with risk, and the obligation of OHRP to provide guidance to IRBs and researchers about appropriate use of different review procedures and to establish a data system to understand and monitor the operations of IRBs.

Because our charge is to consider human research participant protection issues for SBES research, our expertise and our report reflect that charge (e.g., most of the examples we cite are from SBES research or, in some cases, SBES research in a medical setting). Our findings and recommendations, however, have broader application because the boundaries between research domains are not and cannot be sharply drawn. Also, much research today is multidisciplinary—in the composition of the research team, the methods used, and the objectives of the research. Furthermore, both biomedical and SBES research cover the full range of risks to participants: biomedical research includes high-risk clinical trials of experimental drugs, but it also includes much minimal-risk research (e.g., surveys about diet, exercise, or medical treatment and epidemiological studies of the spread of infections); and SBES research also spans the spectrum of risks. However, our analysis and recommendations are usually couched in terms of SBES research or, sometimes, in terms of "research," in keeping with our charge and expertise.

We discuss human participant protection issues for SBES research in the context of the Common Rule (45 *CFR* 46, subpart A), which applies generally to all human participants and has been adopted by most agencies that sponsor research with humans (see Box 1-1).[4] We do not have the expertise and do not address issues of added protections for children, pregnant women, and prisoners, which are covered in other

---

[4]45 *CFR* 46, from which we quote extensively in our report, is the DHHS enactment of the Common Rule; other agencies' enactments are found in other volumes of the *Code of Federal Regulations*.

subparts of 45 *CFR* 46 (see Box 1-2). Infants and children are the subject of much research in such fields as developmental psychology and the sociology of families, and the protection issues for children in SBES research likely merit a full study of their own.

Because SBES covers such a wide variety of fields and topics, we could not hope to examine specific issues in every field to the same extent. For example, we draw somewhat more heavily on studies and data about protection issues in surveys and secondary analyses than in some other fields. Nonetheless, our focus on issues of informed consent, confidentiality protection, and the review process is relevant to and useful for the full range of SBES disciplines.

ACTIVITIES

At its first meeting, our panel received input from the IOM committee and its sponsors, as well as representatives of several professional associations representing the SBES disciplines, ongoing advisory committees, and federal agencies. The panel considered relevant reports of other groups that have given attention to the protection of participants in SBES research, including those of the American Association of University Professors (2001), the Association of American Universities (2000), the National Bioethics Advisory Commission (2001), the DHHS-chartered National Human Research Protections Advisory Committee (NHRPAC),[5] the NHRPAC Behavioral and Social Science Working Group, and the National Science Foundation's ad hoc Social, Behavioral, and Economic Subcommittee for Human Subjects (2002).[6]

The panel reviewed the history of human research participant protection from the perspective of SBES research. It reviewed the few existing studies of the operation of IRBs for relevant information on variation in procedures across IRBs and trends across time, relying particularly on two relatively comprehensive studies, both surveys: one by Bell, Whiton, and Connelly (1998), and one by Cooke, Tannenbaum, and Gray (1978). The panel also examined the IRB guidelines of 47 major research universities, as posted on their Internet web sites, to gain a more up-to-date picture of IRB requirements and guidance with respect to SBES research in large research institutions. Similarly, the panel reviewed human research participant protection guidelines of

---

[5]This committee was replaced by the Secretary's Advisory Committee on Human Research Protections in October 2002; see Appendix B.

[6]We were not able to obtain input directly from participants in SBES research, who cover a wide range of populations or else represent the general population as distinct from a group of patients with a specific disease as is characteristic of much clinical research.

the major SBES professional associations. Finally, the panel brought to bear the expertise of its members, some of whom have served on IRBs or have written in the field of research ethics, and most of whom have conducted research with human participants in a variety of SBES fields.

ORGANIZATION OF REPORT

Our report has seven chapters and five appendices. Chapter 2 reviews basic concepts and background information that is central to our findings and recommendations. Following a summary of principles and practices for human research participant protection that were articulated in the landmark Belmont Report (National Commission for the Protection of Human Subjects of Biomedical and Behavioral Research, 1979), the chapter considers more fully issues of harm, benefit, risk, and minimal risk. It then reviews the available evidence on IRB review of different types of projects and, as context, provides examples of SBES research and issues for participant protection.

Chapter 3 provides historical background, covering the history of federal policies and regulations for human research participant protection in the United States with an emphasis on SBES research.

Chapter 4 covers issues of informed consent and its documentation. It reviews the limited available evidence on how IRBs interpret the federal regulations on consent, recommends research to improve consent procedures and documentation for different types of SBES research and populations studied, and considers such issues as consent procedures for special populations, third-party consent, when signed written consent is unnecessary or inappropriate, and the use of deception in research.

Chapter 5 reviews the history of confidentiality protection for data from participants in SBES research and the changing research environment that increases the threats to confidentiality. It provides recommendations for increased attention to confidentiality protection in ways that do not unnecessarily hinder access to data, particularly for secondary analysis.

Chapter 6 provides recommendations for promoting the use of the flexibility in federal regulations for exempting and expediting minimal-risk research. It also recommends research to build knowledge on risks and harms in SBES research and data collection for better understanding the IRB system and how to enhance procedures for carrying out its oversight function.

Chapter 7 addresses system-level issues that are of particular relevance to the SBES research community. Such issues include ways to facilitate IRB-researcher interaction and the participation of SBES researchers in the development of national policy for human participant protection.

Our report has five appendices: changes in federal regulatory language from 1974 to the present (A); list of organizations and resources for human research participant protection (B); the agenda for the panel's first meeting (C); descriptions of studies of IRB operations, including the panel's review of selected IRB websites of major research universities (D); and a paper commissioned by the panel from George Duncan, "Confidentiality and Data Access Issues for Institutional Review Boards" (E).

The panel's recommendations are addressed to various actors, including IRBs, federal agencies, data archives, and SBES researchers. Several recommendations call for OHRP to provide guidance to IRBs and researchers, in recognition of OHRP's leadership role in the federal system. OHRP responsibilities include not only monitoring the operations of IRBs that review DHHS-funded research, but also providing guidance on human research participant protection for the federal and nonfederal sectors, developing educational programs, and exercising leadership for human participant protection for the U.S. government in cooperation with other federal agencies.[7]

---

[7] See 67 *Federal Register,* 10217, March 6, 2002; see also http://ohrp.osophs.dhhs.gov [4/10/03].

BOX 1-1
Key Features of the Common Rule

The complete text of the Common Rule is in Title 45 *Code of Federal Regulations* (CFR) section 46, subpart A (from which all quoted material below is taken). Every federal agency or department adopting the Common Rule publishes it in a section of the CFR dedicated to that agency.

At present ten departments and seven agencies have adopted the Common Rule by regulation, executive order, or legislation: the Departments of Agriculture; Commerce; Defense; Education; Energy; Health and Human Services (DHHS); Housing and Urban Development; Justice; Transportation; and Veterans Affairs; and the Agency for International Development, the Central Intelligence Agency, the Consumer Product Safety Commission, the Environmental Protection Agency, the National Aeronautics and Space Administration, the National Science Foundation, and the Social Security Administration. The Food and Drug Administration has its own set of human participant protection regulations for research and evaluation of drugs and other products it regulates (21 CFR 50, 56), which are nearly identical to the Common Rule. DHHS and a few other agencies have also adopted additional protections for specific populations of research participants: subparts B, C, and D apply to pregnant women, human fetuses, and neonates; prisoners; and children, respectively. (See Chapter 3 for the history of federal protections for human research participants; see Appendix A for changes in regulatory language from 1974 to the present.)

Applicability

- "This policy [with some exceptions, see "Exempt Research" below] applies to all research involving human subjects conducted, supported or otherwise subject to regulation by any Federal Department or Agency which takes appropriate administrative action to make the policy applicable to such research."

Definitions

- Research—"a systematic investigation, including research development, testing and evaluation, designed to develop or contribute to generalizable knowledge."

- Human subject—"a living individual about whom an investigator (whether professional or student) conducting research obtains (1) data through intervention or interaction with the individual, or (2) identifiable private information."

- Minimal risk—"the probability and magnitude of harm or discomfort anticipated in the research are not greater in and of themselves than those ordinarily encountered in daily life or during the performance of routine physical or psychological examinations or tests."

Assurances

- Process and materials by which a research institution assures the federal government that it will comply with the Common Rule for all research conducted at its site. Institutions must provide a list of IRB members and attest that its IRB(s) will uphold the Common Rule requirements. The government reviews the submission and decides to issue a federal-wide assurance (FWA). (The FWA process replaces a previous multiple project assurance process, which required institutions to submit additional materials, such as IRB procedures.)

BOX 1-1 (continued)

IRBs

- Membership—at least five members with varying backgrounds, including at least one scientist and one nonscientist, and at least one member not affiliated with the research institution.

- Authority—may approve, require modifications to, or disapprove all research covered under the Common Rule at its site (some IRBs review research from more than one institution); may suspend or terminate research that violated IRB requirements or resulted in unexpected serious harm to subjects; no covered research may proceed without IRB approval.

- Operations—must review research at meetings attended by a quorum (or use an expedited procedure, see below); must approve research by a majority of those present, notify investigators in writing of its decision, and re-review approved research at least once a year.

Exempt Research

- Six categories of research are exempt from the Common Rule. They are:

  "(1) Research conducted in established or commonly accepted educational settings, involving normal educational practices, such as (i) research on regular and special education instructional strategies, or (ii) research on the effectiveness of or the comparison among instructional techniques, curricula, or classroom management methods.

  "(2) Research involving the use of educational tests (cognitive, diagnostic, aptitude, achievement), survey procedures, interview procedures or observation of public behavior, unless: (i) information obtained is recorded in such a manner that human subjects can be identified, directly or through identifiers linked to the subjects; and (ii) any disclosure of the human subjects' responses outside the research could reasonably place the subjects at risk of criminal or civil liability or be damaging to the subjects' financial standing, employability, or reputation.

  "(3) Research involving the use of educational tests (cognitive, diagnostic, aptitude, achievement), survey procedures, interview procedures, or observation of public behavior that is not exempt under paragraph (b)(2) of this section, if: (i) the human subjects are elected or appointed public officials or candidates for public office; or (ii) Federal statute(s) require(s) without exception that the confidentiality of the personally identifiable information will be maintained throughout the research and thereafter.

  "(4) Research involving the collection or study of existing data, documents, records, pathological specimens, or diagnostic specimens, if these sources are publicly available or if the information is recorded by the investigator in such a manner that subjects cannot be identified, directly or through identifiers linked to the subjects.

  "(5) Research and demonstration projects which are conducted by or subject to the approval of Department or Agency heads, and which are designed to study, evaluate, or otherwise examine: (i) Public benefit or service programs; (ii) procedures for obtaining benefits or services under those programs; (iii) possible changes in or alternatives to those programs or procedures; or (iv) possible changes in methods or levels of payment for benefits or services under those programs.

BOX 1-1 (continued)

"(6) Taste and food quality evaluation and consumer acceptance studies, (i) if wholesome foods without additives are consumed or (ii) if a food is consumed that contains a food ingredient at or below the level and for a use found to be safe, or agricultural chemical or environmental contaminant at or below the level found to be safe, by the Food and Drug Administration or approved by the Environmental Protection Agency or the Food Safety and Inspection Service of the U.S. Department of Agriculture."

Expedited Review

- IRBs may review certain kinds of minimal-risk research and minor changes in approved research by an expedited procedure (see Box 1-2). This review is done by the IRB chair or one or more members designated by the chair.

Criteria for IRB Approval of Research

- Risks to subjects are minimized.
- Risks are reasonable in relation to anticipated benefits and the importance of knowledge to be gained.
- Selection of subjects is equitable.
- Informed consent will be sought from each subject or his or her legally authorized representative.
- Informed consent will be appropriately documented.
- When appropriate, data collection will be monitored to ensure safety of subjects.
- When appropriate, there are adequate provisions to protect privacy and maintain data confidentiality.

Informed Consent

- Unless waived by an IRB, investigators cannot involve humans in research without obtaining informed consent.
- The information provided in seeking informed consent must include eight elements (e.g., "description of any reasonably foreseeable risks or discomforts to the subject") and may include one or more of six added elements (e.g., "any additional costs to the subject that may result from participation in the research") (see Chapter 4).
- Informed consent elements may be waived under specified circumstances.

Documentation of Consent

- Unless waived, consent must be documented by a written, signed consent form.
- IRBs may issue a waiver when (1) the consent document is the only record linking the participant and the research, and the principal risk is the potential harm from a breach of confidentiality, or (2) the research is minimal risk and involves no procedures for which written consent is normally required outside the research context (e.g., a telephone survey).

BOX 1-2
Categories of Research for Which Minimal-Risk Protocols Can Receive Expedited Review

(1) Clinical studies of drugs and medical devices only when condition (a) or (b) is met.

   (a) Research on drugs for which an investigational new drug application (21 CFR Part 312) is not required. (Note: Research on marketed drugs that significantly increases the risks or decreases the acceptability of the risks associated with the use of the product is not eligible for expedited review.)

   (b) Research on medical devices for which (i) an investigational device exemption application (21 CFR Part 812) is not required; or (ii) the medical device is cleared/approved for marketing and the medical device is being used in accordance with its cleared/approved labeling.

(2) Collection of blood samples by finger stick, heel stick, ear stick, or venipuncture as follows:

   (a) from healthy, nonpregnant adults who weigh at least 110 pounds. For these subjects, the amounts drawn may not exceed 550 ml in an 8 week period and collection may not occur more frequently than 2 times per week; or

   (b) from other adults and children, considering the age, weight, and health of the subjects, the collection procedure, the amount of blood to be collected, and the frequency with which it will be collected. For these subjects, the amount drawn may not exceed the lesser of 50 ml or 3 ml per kg in an 8 week period and collection may not occur more frequently than 2 times per week.

(3) Prospective collection of biological specimens for research purposes by noninvasive means. Examples:

   (a) Hair and nail clippings in a nondisfiguring manner;

   (b) deciduous teeth at time of exfoliation or if routine patient care indicates a need for extraction;

   (c) permanent teeth if routine patient care indicates a need for extraction;

   (d) excreta and external secretions (including sweat);

   (e) uncannulated saliva collected either in an unstimulated fashion or stimulated by chewing gumbase or wax or by applying a dilute citric solution to the tongue;

   (f) placenta removed at delivery;

   (g) amniotic fluid obtained at the time of rupture of the membrane prior to or during labor;

   (h) supra- and subgingival dental plaque and calculus, provided the collection procedure is not more invasive than routine prophylactic scaling of the teeth and the process is accomplished in accordance with accepted prophylactic techniques;

   (i) mucosal and skin cells collected by buccal scraping or swab, skin swab, or mouth washings;

   (j) sputum collected after saline mist nebulization.

BOX 1-2 (continued)

(4) Collection of data through noninvasive procedures (not involving general anesthesia or sedation) routinely employed in clinical practice, excluding procedures involving x-rays or microwaves. Where medical devices are employed, they must be cleared/approved for marketing. (Studies intended to evaluate the safety and effectiveness of the medical device are not generally eligible for expedited review, including studies of cleared medical devices for new indications.) Examples:

(a) Physical sensors that are applied either to the surface of the body or at a distance and do not involve input of significant amounts of energy into the subject or an invasion of the subject's privacy;

(b) weighing or testing sensory acuity;

(c) magnetic resonance imaging;

(d) electrocardiography, electroencephalography, thermography, detection of naturally occurring radioactivity, electroretinography, ultrasound, diagnostic infrared imaging, doppler blood flow, and echocardiography;

(e) moderate exercise, muscular strength testing, body composition assessment, and flexibility testing where appropriate given the age, weight, and health of the individual.

(5) Research involving materials (data, documents, records, or specimens) that have been collected or will be collected solely for nonresearch purposes (such as medical treatment or diagnosis). (Note: Some research in this category may be exempt from the HHS regulations for the protection of human subjects. 45 CFR 46.101(b)(4). This listing refers only to research that is not exempt.)

(6) Collection of data from voice, video, digital, or image recordings made for research purposes.

(7) Research on individual or group characteristics or behavior (including, but not limited to, research on perception, cognition, motivation, identity, language, communication, cultural beliefs or practices, and social behavior) or research employing survey, interview, oral history, focus group, program evaluation, human factors evaluation, or quality assurance methodologies. (Note: Some research in this category may be exempt from the HHS regulations for the protection of human subjects 45 CFR 46.101 (b)(2) and (b)(3). This listing refers only to research that is not exempt.)

(8) Continuing review of research previously approved by the convened IRB as follows:

(a) Where (i) the research is permanently closed to the enrollment of new subjects; (ii) all subjects have completed all research-related interventions; and (iii) the research remains active only for long-term follow-up of subjects; or

(b) Where no subjects have been enrolled and no additional risks have been identified; or

(c) Where the remaining research activities are limited to data analysis.

(9) Continuing review of research, not conducted under an investigational new drug application or investigational device exemption where categories two (2) through eight (8) do not apply but the IRB has determined and documented at a convened meeting that the research involves no greater than minimal risk and no additional risks have been identified.

SOURCE: Verbatim quotes from 63 *Federal Register* 60364-60367 (November 9, 1998).

— 2 —
Basic Concepts

IN THIS CHAPTER we introduce two overarching themes that are critical for our findings and recommendations. First is the need for continued vigilance by all those involved in the U.S. human research participant protection system—researchers, institutional review boards (IRBs), research institutions, funding agencies, and the Office for Human Research Protections (OHRP)—to maintain the principles for participant protection that were articulated in the Belmont Report produced by the National Commission for the Protection of Human Subjects of Biomedical and Behavioral Research (1979). Second is the need to maintain that vigilance in a way that is commensurate with the risk of each research protocol. Following a summary of the Belmont Report principles and practices that follow from them, the chapter considers more fully issues of harm, benefit, risk, and minimal risk. It then considers the current mismatch between the risks of research projects and the type of review afforded them by many IRBs. Finally, as context, the chapter discusses examples of social, behavioral, and economic sciences (SBES) research and issues for participant protection.

PRINCIPLES AND PRACTICES FOR ETHICAL RESEARCH

General Principles

Although U.S. policies and regulations for protection of human research participants date back to the 1960s (see Chapter 3), basic ethical principles underlying and informing such regulations were not articulated until 1979, when the national commission issued the Belmont Report. That report identified three major ethical principles for the conduct of research on humans—respect for persons, beneficence, and justice:

> *Respect for Persons*—the obligation to treat individuals as autonomous agents whose decisions on whether or not to participate in research are to be respected and not overridden by a researcher. From this principle follows the requirement for researchers to obtain voluntary informed consent

23

from participants. Special recognition must be given to issues of respect when dealing with people who are immature, incapacitated, or whose autonomy is constrained. Those with limited capacities need to be protected from harm by providing for consent by authorized proxies and by taking extra care to minimize research risks or, in some cases, precluding their participation in research.

Beneficence—the obligation to secure participants' well-being by protecting them from harm to the extent possible and by maximizing the benefits—to them especially, but also to society—that are expected to result from the research. From this principle follows the requirement for researchers and IRBs to assess risks of harm and probability of benefits in a systematic manner.

Justice—the obligation to show fairness in the selection of research participants with regard to the distributions of the burdens and benefits of the research. From this principle follows the requirement for researchers to select participants in an equitable manner for particular studies and for funding agencies to consider the distribution of burdens and benefits across society (e.g., to ensure that certain groups are not systematically excluded from or included in research).

Applying Principles to the Conduct of Research

Everyone concerned with research on humans should be fully cognizant of the Belmont principles in designing and reviewing protocols and monitoring ongoing research. Resolving conflicts among principles, however, can prove challenging in practice and underscores the necessity of the ethical review processes that are in place for research with humans. In practice, the three principles translate into consideration of three requirements: informed consent, assessment and appropriate balancing of risks and benefits, and fair procedures for selection of research participants. In addition, although not explicitly articulated in the Belmont Report, the principles support the protection of confidentiality. (See also Box 1-1 in Chapter 1, which lists the criteria that IRBs must consider in reviewing research protocols.)

Informed Consent—providing an individual with comprehensible information regarding known risks of harm, possible benefits, and other details of the proposed study prior to the point at which the person freely chooses to participate. By providing full information to prospective participants, researchers assure that each of them can decide whether he or she is willing to participate given his or her situation and personal tolerance for risk. Consider, for example, a test of an experimental drug for the treatment of a mental illness when the drug is known to have a number of potentially serious side effects. A less invasive example would be a psychological experiment in which a lengthy series of mental tests are administered to elderly persons over the course of a few hours. In the second case some temporary fatigue or distress is likely, which may be regarded as harmful to some people. Regardless of an experimenter's belief in the potential benefits to the participant or the long-term benefits from the research, it would be unethical for the experimenter to subject the person to these kinds of risks without consent.

The right to decide about participation on the basis of full information is not limited to studies that pose significant risks of harm. It exists for studies that are as inconsequential as stating color preferences for automobiles in market research, as well as for studies probing the effect of grieving on the emotional health of a surviving spouse. Under carefully considered circumstances, however, it can be appropriate to use less than fully informed consent—for example, keeping information about a particular feature of a study from a prospective participant until the study is completed when such information would likely alter the participant's behavior, the knowledge to be gained is important, and the risk to the participant from omitting the particular information is minimal.

Assessment of Harms, Risks, and Benefits—weighing and appropriately balancing the risks of harm and the potential for benefits from participation in the proposed study. Although there is little disagreement about the desirability of minimizing harm and maximizing benefits from participation in research, determining for a specific research protocol the type and extent of harm, the probability or likelihood of harm, and the benefits likely to be obtained from participation is, at best, inexact. Such assessments are almost always subjective and often involve issues on which reasonable people disagree. Yet such judgments cannot be avoided (see "Harms, Risks, and Benefits" below).

Fair Selection of Research Participants—assuring fair procedures and outcomes in the selection of research participants. Achieving fairness requires consideration of those who are included in research and those who are excluded. If participation is believed to be beneficial to either the participants or the populations represented by them, then excluding some people raises an issue of fairness. For example, early studies of cardiovascular disease rarely included women, leading to knowledge with potential limitations for understanding cardiovascular diseases in women. If participation is believed to carry significant risks of harm, then restricting research to particular population groups is also an issue of fairness, particularly if those groups are subject to coercion (e.g., prisoners who are denied privileges or offered added privileges to participate).

Confidentiality Protection—keeping the participant's identity confidential. Confidentiality is another means of showing respect for a person. A person has the right to expect that, if he or she participates in research under conditions of confidentiality, the researcher will respect and assure that confidentiality. Confidentiality may also address beneficence. In some cases, making research information public could put a participant at risk. For example, if sensitive personal information became known to the person's employer, it could put his or her job or benefits in jeopardy.

HARMS, RISKS, AND BENEFITS

In this section we briefly discuss some of the critical factors surrounding the judgments about harms, risks, and benefits that are necessary to address the ethical principle of beneficence.

Types of Harm

Drawing on the final report of the National Bioethics Advisory Commission (NBAC) (2001:71-72) and adding examples from SBES research, below we discuss six types of harms that can occur to research participants: physical, psychological, social, economic, legal, and dignitary.[1]

- **Physical harm** from research can include death, injury, pain, suffering, or discomfort. Examples in biomedical research range from death due to an experimental drug administered in a cancer study to discomfort from having to keep still for a long time during an MRI (magnetic resonance imaging) study. Examples in SBES research range from death or injury due to the failure of an alternative automated method of helping blind people cross at traffic signals to discomfort from being subjected to loud noises or bright lights during a stimulus-response study. Physical harm, including injury and death, can also result from a breach of confidentiality that discloses sensitive information (e.g., that one is participating in a study of gang violence, which could lead to retaliation by gang members).

- **Psychological harm** from research can include negative self-perception, emotional suffering (e.g., anxiety or shame), or aberrations in thought or behavior (e.g., agreeing to a hateful statement under pressure from the research environment). In both biomedical and SBES research, psychological harm from the research procedure can range from momentary anxiety or embarrassment to long-lasting, intense psychological distress and fear, which could in extreme cases result in suicide. A biomedical example is when a participant in a genetics study learns that he or she is likely to develop a disease for which there is no treatment or cure. An SBES example is when a participant in a study on traumatic events recalls memories that are intensely distressing. Psychological harm, such as distress, anger, or guilt, can also result from disclosure of sensitive or embarrassing information collected in the research.

- **Social harm** can involve negative effects on relationships or interactions with other people. Such effects are most likely to result from a breach of confidentiality, in which a participant's answers become known to others. Examples of social harm include discriminatory behavior resulting in loss of insurance or employment from knowledge of study results (e.g., that one has or is likely to contract a specific disease). Stigmatization is another

---

[1] Recent guidance from the National Science Foundation (2002:17-18) is similar but omits dignitary harm and includes "moral harm when participation in research strengthens the subjects' inclinations to behave unethically."

social harm that can result from knowledge about a person's participation in a study or particular findings.

- *Economic harm* usually involves financial loss, which can result from study participation (e.g., the need to pay for transportation or child care in order to participate), from disclosure of study findings or participation (e.g., loss of health insurance or employment), or as a side effect of other harms (e.g., having to pay court costs in a lawsuit that results from a breach of confidentiality).

- *Legal harm* can include arrest, conviction, incarceration, and civil lawsuits. Such harm can result, for example, from a breach of confidentiality in studies of possession or use of illegal drugs, sexual abuse, or shoplifting behavior, or in situations in which state law requires that certain types of researchers report particular activities, such as child abuse.

- *Dignitary harm* can result when individuals are treated as means to an end and not as people deserving respect for their own values and preferences. Such harm can happen in studies that do not appropriately obtain informed consent.

Research projects can pose risks of more than one type of harm (e.g., stigmatization, psychological stress, and financial loss from disclosure of confidential information). Research projects can also result in harm to people not directly involved in the research (see National Bioethics Advisory Commission, 2001:72). For example, family members could be stigmatized or otherwise harmed by a breach of confidentiality that disclosed information about a family from an individual's participation in genetic research. Figure 2-1 shows a distribution of the kind(s) of harm that a sample of investigators of biomedical and SBES research projects anticipated could potentially result to participants in their projects, with a slightly different categorization than we use.

Differences in methods used in SBES research relate to the appropriate focus of IRBs in determining the kinds of potential harm to human participants. For research involving interventions, such as a laboratory experiment in which the participant is subjected to a stimulus, a primary focus must be on the potential harm to the participant from the intervention itself. The potential harm from a breach of confidentiality is also of concern. For research in which the participant is answering questions from a researcher, the primary focus is on the harm that could result from a breach of confidentiality. The psychological harm from asking sensitive questions is also of concern and is affected by

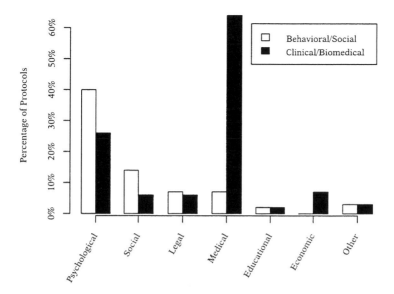

Figure 2-1 Types of Possible Harm Anticipated by Investigators for Protocols, by Type of Research

NOTE: Classification by investigators (n = 632): behavioral/social research includes social science, behavioral science, educational research, and health services research; clinical/biomedical research includes clinical research, biomedical science, and epidemiology.

SOURCE: 1995 survey of IRBs in Bell, Whiton, and Connelly (1998:Figure 11a).

whether the researcher assures the participant that any such question can be skipped. For research that involves no contact between the researcher (or research team) and the participant, the primary concern is the potential harm from a breach of confidentiality.

Procedures to encourage participation also raise the potential for harm. For experiments, one problem may arise when volunteers become so motivated by direct incentives to participate, such as the possibility that they or a close relative or friend will benefit from an experimental treatment, that they fail to take adequate account of the risks of participation. Another problem can occur if volunteers are so coerced (overtly or in subtle ways) that their right to voluntarily participate is not respected (e.g., students who perceive that participating in an experiment is necessary to remain on good terms with the instructor). Yet another problem can arise for surveys for which achieving high response rates among randomly selected (presumably disinterested) people is a key issue for the quality of the research results: What are the appropriate procedures to ensure participation without harming participants by the recruitment procedure (e.g., by making their identity known to others)?

Estimating Risks of Harm

It is generally not difficult to imagine types of harm that particular research projects may pose. What is often difficult to estimate accurately is the severity of the harm and the likelihood that it will occur—that is, to estimate risk. It is particularly difficult to estimate risk for many types of nonphysical harms given the absence of a good base of evidence. As the National Bioethics Advisory Commission (2001:72) notes:

> Determinations concerning the probability of physical harms are often easier to make than those involving the probability of nonphysical harms. For example, the magnitude and probability of harms associated with a blood draw are well known and can be objectively quantified. This is generally not the case for psychological, social, economic, and legal harms. ... IRBs, therefore, can err *in either direction* [italics added], by assuming a higher probability and recommending unnecessary protections or preventing research from being conducted or by assuming a lower probability and allowing research to occur without all the appropriate protections. ... [Also,] although a good deal of information has been gathered about some nonphysical harms—for example, the risks from disclosures associated with transmitting or storing certain types of information—the possibility of such harms is not widely appreciated.

Assessments of the extent to which IRBs overestimate (or underestimate) risks of different types of harms are limited (see "Role of IRBs" below). Moreover, even if IRBs and researchers agree on the risks of a particular research study, it may still be a matter of judgment as to whether the study meets the Common Rule definition of posing no more than "minimal risk" to participants (see "Minimal Risk" below).

Benefits

Benefits can be as difficult to identify and quantify as the risks of harm. Balancing risks of harm against likely benefits, particularly when the benefits are indirect, is also far from easy. For experimental biomedical research, benefits are often thought of as improved medical treatments for illnesses or disabilities. Yet a major issue for clinical trials of experimental drugs or devices is that participants may confuse research with medical care and expect an immediate benefit to themselves when such benefit may not be likely even if the participant receives the experimental treatment and not a placebo.

For most, if not all, SBES research, there is usually little direct benefit to participants in the sense that the results of the research will be of immediate help to them, but SBES and biomedical research can provide two other kinds of benefits. The first type of benefit is when knowledge about humans and human societies helps decision making in the public and private sectors by individuals, households, businesses, organizations, and governments. For example, from psychological research much has been learned about the human brain and the kinds of stimuli that are essential to the development of cognitive, social, and emotional skills. This knowledge has been used by parents, educators, and others to help children grow. From economic decision-making research, knowledge has been gained about how people respond to financial incentives. This knowledge has been used to craft policies to encourage saving. From survey research have come indicators of consumer spending and confidence in the economy that are important forecasters of economic growth or recession.

A second type of benefit in much SBES and biomedical research comes from the study procedure. This type includes such benefits as the opportunity for education and gaining access to information (e.g., information about nutrient contents of foods in a study of food-buying patterns of low-income families or resources for child-rearing advice in a study of mother-child interactions) and the opportunity to earn the esteem of other participants and the research team. These kinds of benefits can be meaningful to participants and help build positive long-term relationships with a research program (see Sieber, 1992:Ch. 9).

MINIMAL RISK

Driven primarily by the nature of the IRB process, a normative "minimal-risk" construct has evolved. It plays a central role in the sequential decisions by IRBs regarding the type of review for each protocol. If the protocol involves research with human participants (see Chapter 6), the first decision is whether the IRB will exempt the protocol from review. If the first decision is to review the protocol, the next decision is whether the IRB will conduct an expedited review or a full committee review. The latter is required when the protocol is not eligible for exemption and the IRB determines that it involves more than minimal risk. Having determined the type of review, the IRB then must conduct that review to evaluate the research practices and procedures of the protocol as they relate to the ethical treatment of human participants, including judgments about the key practices discussed earlier—informed consent, balancing of risks and benefits, selection, and confidentiality—considering both the vulnerability of the population of interest and who is being invited to participate in the study. IRBs must impose stringent requirements for informed consent when the IRB judges a protocol to be more than minimal risk.

The Common Rule (45 CFR 36.102i) defines "minimal risk" to mean that "the probability and magnitude of harm or discomfort anticipated in the research are not greater in and of themselves than those ordinarily encountered in daily life or during the performance of routine physical or psychological examinations or tests." Beyond that definition, little concrete guidance is available to IRBs for determining minimal risk.

Moreover, the definition itself is ambiguous in several respects. For example, a "routine" psychological test may be of more than minimal risk when it is performed on severely depressed people. Furthermore, different populations experience different risks in daily life—for example, the risks that combat soldiers willingly accept as part of training are much greater than the risks that white-collar workers would accept as part of their jobs. Also, some populations (e.g., poor children in bad neighborhoods) experience high levels of risk in their daily lives through no choice of their own.

Not surprisingly, views differ on what constitutes minimal risk. The National Human Research Protections Advisory Committee Social and Behavioral Science Working Group recently attempted to define minimal risk as meaning "that the worst harm that could occur in a study should not be very serious—even if many subjects experience it, and, if the harm is serious, then the probability of any given subject experiencing it should be quite low."[2] This formulation suggests not only that projects posing no or minor harm to participants and having a low probability that harm will occur are minimal risk, but also that projects posing no or minor harm to participants and having a high-probability that harm will occur are minimal risk. Recent guidance from the National Science Foundation (2002:9) agrees, noting, in par-

---

[2]This is a draft statement; see http://www.asanet.org/public/humanresearch/riskharm02.html [4/10/03].

ticular, that a high probability harm can be minimal risk provided that the magnitude of the harm is very low. An example is an innocuous survey that annoys the respondent by taking longer than he or she would like. Even if most or all respondents are annoyed, an innocuous survey is still minimal risk because the harm to any one respondent is minor and fleeting, and people experience similar transitory annoyances every day.

In addition, the working group formulation suggests that projects posing serious harms to participants can be minimal risk if the probability of such harm occurring to any given participant is extremely low. Barnbaum (2002), however, argues that such projects should not be treated as minimal risk because serious harm could occur for one or more participants. For example, a police officer who participated in a study of police officers' views on police corruption and violence could lose his or her job if confidentiality were breached and his or her participation disclosed.

We agree that the example cited by Barnbaum should not be treated as minimal risk. However, just because a serious harm can be imagined does not mean that a project must be treated as more than minimal risk. In a survey of the general population, it is almost always possible to imagine that some respondent somewhere could have a negative reaction to being questioned that could, theoretically, result in a serious harm, such as a relapse for a person suffering from depression. However, such relapses may occur for many reasons in the course of daily life. If adequate measures are taken to inform prospective respondents of the nature of the survey and their right not to answer some or all questions, then the mere possibility that a random respondent might have an adverse reaction should not be sufficient reason to take the project out of the minimal-risk category. For that to occur, there should be evidence that particular questions have had significant adverse effects, or there should be a direct link of the possible harm to the type of respondent, as in the case of the police officer example.[3]

We further believe that, when determining the level of risk, it is important to consider not only the possible intensity of the harm, but also its likely duration. For example, the occurrence of psychological harm in a research project could result in one of three situations: (1) a minimal and fleeting annoyance or other emotion; (2) a sharp but short-lived feeling of anxiety, embarrassment, anger, or other emotion; or (3) an intense and long-lasting feeling of anxiety, anger, guilt, or other strong emotion. Of these three situations, we argue that the second as

---

[3] In Chapter 6 we discuss the need for SBES researchers to document harm to participants as a means to build an evidence base; see also Chapter 7 for a discussion of the desirability of systematic research on risks and harm of different kinds of research.

well as the first is most often minimal risk. Only the third situation seems a situation of greater than minimal risk.

Another issue in connection with minimal risk is the standard of comparison when evaluating the risks of the research against the risks of daily life: Whose daily life is to be the comparison? Federal regulations use a high standard for research on prisoners, namely, that minimal risk is that of nonincarcerated healthy individuals (45 *CFR* 46, subpart C). The Office for Protection from Research Risks of the National Institutes of Health (NIH) endorsed that same high standard for research with the general population in 1993. However, NBAC argued that such a high standard for the Common Rule (45 *CFR* 46, subpart A) goes against the history of human participant protection regulation. For example, the preamble to the 1981 version of 45 *CFR* 46 stated that "the risks of harm ordinarily encountered in daily life means those risks encountered in the daily lives of the subjects of the research" (46 *Federal Register* 8366; see also Appendix A).

However, NBAC does support an interpretation in which the standard for minimal risk is the general population. Such a standard, while not as restrictive as one using healthy individuals, is more restrictive than a relative standard, in which risks are defined relative to the particular research population. For example, a relative standard might say that bone marrow aspiration is minimal risk for people with acute leukemia, but a general population standard would classify such a procedure as more than minimal risk (National Bioethics Advisory Commission, 2001:83).[4]

We are not prepared to reach a conclusion about the appropriate population standard for minimal risk. We believe that the issue merits wide debate that will, hopefully, lead to useful guidance for IRBs and researchers. Such debate should involve not only the small circle of ethicists who have considered the matter, but also the broader community of IRB members, researchers, and representatives of participants.

We argue in subsequent chapters that much more concrete guidance is needed for IRBs and researchers on the kinds of research protocols that qualify as minimal risk. We also acknowledge that there will always be a role for judgment on the part of IRB members to apply appropriately the Common Rule regulations and guidance regarding minimal risk to individual research populations and settings.

---

[4]It is not clear whether a "general population" standard would refer only to the U.S. population or how an evaluation of minimal risk should be applied to research that involves participants from other countries.

ROLE OF IRBS

Consideration of minimal risk leads to a consideration of the functioning of IRBs because the minimal-risk construct plays such a prominent role in the decision making of individual IRBs as they deal with individual research protocols. When considering the overall decision making represented by the total set of IRB judgments on all protocols, two major criticisms of the current IRB system arise: (1) that IRBs are overloaded, underfunded, and, consequently, hard pressed to fully carry out their responsibilities for protecting human participants in more-than-minimal-risk research; and (2) that IRBs are spending too much time on scrutinizing minimal-risk research (perhaps as a reaction to heightened scrutiny of IRB operations by the federal government and the media in the light of highly publicized deaths to research participants; see Chapter 3). To the extent that overreview of minimal-risk research is interfering with the ability of IRBs to properly review higher risk research, then these two criticisms are two sides of the same coin—namely, the problem of determining the risk in a research protocol and acting appropriately on that determination.

Comparative, reliable data on the operations of IRBs are scarce, an issue that we address later in Chapter 6. However, we believe the available information is sufficient to warrant three conclusions: (1) IRBs are indeed overburdened; (2) IRB practices regarding the type of review vary considerably across IRBs; and (3) this variability is much more likely to affect the type and nature of review afforded minimal-risk research compared with research that is of more than minimal risk. These findings imply that the resources of many IRBs are not being used as effectively as they could be and that standards for reviewing research have a sizable idiosyncratic element across IRBs.

The Institute of Medicine (IOM) committee recommends increased resources for IRBs (see Chapter 7). We agree but add that using these resources simply to devote more time and energy to reviewing protocols may not be sufficient. Such resources should also be invested in aiding the development and application of consistent guidelines for types of review that are commensurate with risk. Having such guidelines is likely to reduce workloads that result from using inappropriate procedures for review of minimal-risk research. To help develop guidance for risk determination and the application of types of review, it is incumbent upon researchers to develop a knowledge base about the risks and harms that are likely for different kinds of research and about appropriate informed consent procedures and related topics. Such knowledge can inform OHRP guidance, assist IRB decision making, and contribute to improved understanding among researchers about ethically responsible research designs.

Heavy Workload

There has been significant growth in IRB workloads over time. It appears that at least half of IRBs at academic research institutions have heavy workloads, with the number of reviews per year (including initial reviews of new projects, continuing reviews, and reviews of proposed changes to previously approved projects) totaling more than the number of calendar days. More specifically, the available evidence shows the following:

- In 1975, IRBs averaged 43 initial reviews per year; by 1983, the average number of initial reviews had increased to 133 per year. In 1995 (the latest available data), the average number of initial reviews had increased to 214 per year. In 1995, the average number of all reviews (initial, continuing, and changes) totaled 578 per IRB.[5]

- The average number of all reviews in 1995 varied from 87 reviews for IRBs in the lowest 10 percent of IRB workloads to 2,144 reviews for IRBs in the highest 10 percent of IRB workloads. IRBs in the top half of the distribution averaged more reviews than days in the year (see Figure 2-2; computed from Bell, Whiton, and Connelly, 1998:7,9).

- Because high-volume IRBs had such heavy workloads, they accounted for a disproportionate share of reviews: those IRBs in the highest decile of IRB workloads accounted for 37 percent of the total estimated number of reviews; those IRBs in the highest 50 percent of IRB workloads accounted for 88 percent of the total estimated number of reviews.

To handle their heavy workloads, high-volume IRBs (those in the highest 10 percent of IRB workloads) function differently than low-volume IRBs (those in the lowest 10 percent of IRB workloads). Based

---

[5]The 1975 data are from a study by the University of Michigan for the National Commission for the Protection of Human Subjects in Biomedical and Behavioral Research (Cooke, Tannenbaum, and Gray, 1978; Gray, Cooke, and Tannenbaum, 1978; hereafter the 1975 survey); the 1983 data are from Grundner (1983); the 1995 data are from a study commissioned by the NIH Office for Extramural Research (Bell, Whiton, and Connelly, 1998; hereafter the 1995 survey). The 1975 and 1995 surveys represent IRBs at research institutions with multiple project assurances or the equivalent (see Appendix D); very little information is available about IRBs in other settings, such as community-based hospitals (see "IRBs with Very Low Volume" below).

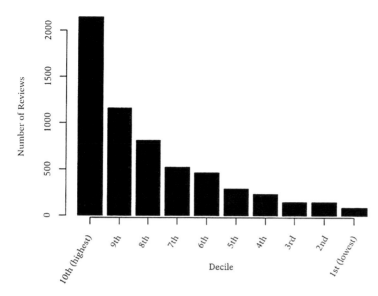

Figure 2-2 Average Reviews by IRBs in Each Decile of Workload Volume, 1995

NOTE: Reviews include initial reviews, continuing or annual reviews, and amendments to approved protocols; data provided by IRB chairs (n = 394) for 1995 or the most recently completed year of record. Workload volume deciles computed by Bell, Whiton, and Connelly on the basis of initial reviews only.

SOURCE: Computed from Bell, Whiton, and Connelly (1998:7,9).

on average data from the 1995 survey (Bell, Whiton, and Connelly, 1998:Ch.IV):

- High-volume IRBs met more often than did low-volume IRBs (21 meetings and 5 meetings per year, respectively).

- Chairs of high-volume IRBs spent more time on all IRB activities, including meetings, preparation, and other activities, than did chairs of low-volume IRBs (386 hours and 72 hours per year, respectively).

- Members of high-volume IRBs spent more time on all IRB activities than did members of low-volume IRBs (108 hours and 28 hours per year, respectively).

- High-volume IRBs had more members than low-volume IRBs (20 members and 10 members, respectively).

- All high-volume IRBs had administrative staff support (122 hours per month); only 15 percent of low-volume IRBs had such support (12 hours per month over all low-volume IRBs).

- High-volume IRBs were more likely to use consultants to review individual proposals than were low-volume IRBs (a median of 33 times per 100 initial reviews and 1.4 times per 100 initial reviews, respectively).

Yet despite these coping strategies, high-volume IRBs spent less time than low-volume IRBs on review of individual projects. The average times from the 1995 survey show striking differences:

- High-volume IRBs spent less full board meeting time (3 minutes) on initial proposal reviews than did low-volume IRBs (21 minutes). (Full board meeting time covered more-than-minimal-risk projects and such minimal-risk projects as the IRB decided not to exempt or review with an expedited procedure.)

- High-volume IRBs spent less total time on review of all initial proposals (7 hours) than did low-volume IRBs (15 hours), including time spent on meetings, preparation, recording and review of minutes, etc., by IRB chairs, members, administrators, and staff.

Over time, the trend is toward substantially less time spent by IRBs on initial proposal review:

- Full board meeting time averaged only 8 minutes per initial proposal review in 1995, compared with 1 hour in 1975.

- Total time spent per review of all initial proposals averaged 7 hours for high-volume IRBs in 1995 and 15 hours for low-volume IRBs in 1995, compared with 38 hours for all IRBs in 1975 (Bell, Whiton, and Connelly, 1998:48,51; Gray, Cooke, and Tannenbaum, 1978:1095).[6]

Clearly, IRBs are stretched thin. Whether that situation adversely affects human participants is not an easy matter to assess. Reports of harm to research participants are not compiled and made available in ways that would help answer the question, and some harm may not be reported. The media have been assiduous in publicizing unexpected deaths of research participants (see Chapter 3), but the numbers, severity, and duration of injuries or other types of harm have not been documented in any systematic way (see Institute of Medicine, 2002:Ch.5).

---

[6]There was no provision for exemption or expedited review in 1975.

Even with good data on harm, it would not necessarily be clear the extent to which deficiencies of IRB review played a causal role rather than other factors, such as failure by researchers to carry out the research as proposed or just misfortune.

Yet the data on the limited amount of time spent per initial review among high-volume IRBs and the very limited time spent on full board review by such IRBs does suggest that the IRBs that handle most of the research workload may not be well positioned to identify potential risks of harm to human participants. This may be true even if high-volume IRBs operate more efficiently than low-volume IRBs. Perhaps supporting such a conclusion is a finding in the 1995 Bell survey that most of the changes IRBs required investigators to make to their proposals in order to gain approval dealt with the form used to document consent (Bell, Whiton, and Connelly, 1998:Figure 41). Compared with 78 percent of proposals for which the consent form had to be modified, investigators were much less often asked to change other aspects of their studies: consent procedure, 21 percent; privacy or confidentiality protection, 14 percent; participant recruitment, 11 percent; scientific design, 6 percent; all other areas combined, 27 percent.[7]

Inappropriate Level of Review

In the perception of the SBES research community, IRBs often overestimate the risks of SBES protocols, resulting in the application of Common Rule provisions that were developed for more-than-minimal-risk research to minimal-risk research. Moreover, because some but not all IRBs appear to use more stringent review standards than the regulations or the risk level of many protocols require, researchers face varying standards for review when they are involved in multi-institution projects or move from one research institution to another.

In spite of anecdotal concerns raised about IRB behavior, there is little hard evidence about the extent to which IRBs may be overestimating risks of the protocols they review, particularly with regard to the minimal-risk versus more-than-minimal-risk distinction in the Common Rule. We found only one study of human research participant protection that included an independent assessment of risks. In that study, members and staff of the Advisory Committee on Human Radiation Experiments (1996:443) reviewed 125 biomedical research projects funded in the early 1990s (mainly radiation studies). They

---

[7]The 1975 Michigan survey found that the attention that IRBs focused on consent forms was not productive in that it did not result in more complete or readable forms (see Chapter 4).

classified 60 percent as minimal risk or perhaps minimal risk and the remaining 40 percent as clearly more than minimal risk. The study did not report on the extent to which the study assessments about risk agreed with assessments of either the IRBs or the researchers involved. While reports by investigators are consistent with a conclusion that significant percentages of research are minimal risk, consistency is not strong support.[8] Investigators' reports may be biased toward underestimating types of harms and levels of risk, and IRBs may not agree with an investigator's viewpoint. Given the obvious advantage of accurate risk assessment for human participant protection, the effective functioning of IRBs, and the credibility of the oversight process in the eyes of participants and researchers, further research on risk determination is certainly warranted.

There is more evidence about the extent to which IRBs are not using the flexibility in the Common Rule that permits less than full board review for research that the IRB itself agrees is minimal risk. This flexibility dates back to 1981 when the Common Rule (then a regulation only of the U.S. Department of Health and Human Services) specified several categories of research that IRBs could exempt from review and additional categories of minimal-risk research that IRBs could review by an expedited procedure in which the chair or a subcommittee of board members would conduct the review instead of the full board. The explicit intent of these provisions, which were implemented after a major battle involving the SBES research community (see Chapter 3), was to exempt a large proportion of minimal-risk research (much of it SBES research) and to allow IRBs to use an expedited procedure for review of many other projects that were deemed to be minimal risk.[9]

Exemption

A 1983 study found reluctance among IRBs to avail themselves of the new Common Rule exemption provisions: almost all IRBs at that time had decided not to exempt research projects from review that fell under one of the four eligible categories of educational, social, and be-

---

[8]The 1995 Bell survey reported that three-fourths of projects were judged by investigators to have less than 10 percent likelihood of a "low" degree of harm (Bell, Whiton, and Connelly, 1998:20). Similarly, the 1975 Michigan survey reported that one-half of projects were judged by investigators to be without risk or to have a "very low" probability of "minor" medical or psychological complications. Risk assessments were obtained for over 2,000 projects (Gray, Cooke, and Tannenbaum, 1978:1096-1097).

[9]Exemption does not require explicit determination of minimal risk, but the categories are designed to exempt SBES (and biomedical) research that is minimal risk (e.g., because no identifying information is obtained), as well as SBES research that involves public officials or programs.

havioral research (Grundner, 1983). Even by 1995, when six categories of research could be exempted, 48-63 percent of IRB chairs reported that their standard practice was *not* to exempt research that fell into one of the categories (e.g., 60% of chairs reported not exempting research using tests, surveys, or observations as standard practice).[10] Furthermore, 35 percent of IRBs reported that they *never* exempted any research from review (Bell, Whiton, and Connelly, 1998:9,29). We do not know how these percentages may have changed across all IRBs since 1995. Our review of IRB websites of 47 major research universities in late 2002 found that relatively few IRBs at these institutions—9 percent—did not offer an option to exempt research from review. We also do not know how many IRBs operated at the other extreme—that is, always granting an exemption requested by an investigator.

Expedited Review

As of 1995, many IRBs did not use the option to expedite the review of minimal-risk projects that fell under one of the specified categories but, instead, gave many such projects full board review. Thus, for three SBES-related categories—existing data, voice recordings, and individual or group behavior—42 percent, 50 percent, and 51 percent of IRB chairs, respectively, reported that their standard practice was full board review.[11] Moreover, 15 percent of IRBs reported that they never expedited any initial reviews (Bell, Whiton, and Connelly, 1998:10,30). Our review of IRB websites of 47 major research universities in late 2002 produced a similar finding—13 percent of IRBs at these institutions did not offer an option for expedited review. At the other extreme, 2 percent of IRBs in the 1995 survey conducted no full board reviews— that is, all new protocols were reviewed by an expedited procedure or were exempted from review.

Variability

At present, there appears to be wide variability in the extent to which IRBs avail themselves of the option for either exemption or expedited review. Moreover, such variability is not linked to IRB workload: high-volume IRBs are not significantly more likely than low-volume IRBs to use the provisions for exemption and expedited review.

---

[10]These percentages are averaged across high-volume and low-volume IRBs and are very similar for the two groups.

[11]The classification of research categories eligible for expedited review differed somewhat at the time of the 1995 survey from the current list (see Box A-5 in Appendix A). These percentages are averaged across high-volume and low-volume IRBs and are very similar for the two groups.

Thus, in 1995 the distribution of exempt protocols, initial expedited reviews, and full board reviews as percentages of all initial reviews— 15 percent, 26 percent, and 59 percent, respectively, over all IRBs— hardly varied across workload-volume deciles (Bell, Whiton, and Connelly, 1998:9-10). However, IRBs within each workload-volume decile exhibited extreme variability: every decile included one or more IRBs for which more than 95 percent, or less than 5 percent, of their workload comprised full board reviews, with the percentages of full board reviews for the remaining IRBs spread out fairly evenly between these two extremes.

Burden

From these data, it is clear that many IRBs are not exempting or expediting as much research as they could under the Common Rule provisions. Because of substantial differences in estimated time spent by type of review, these IRBs are therefore adding to their review burden and that of investigators in preparing for review. Investigators in the 1995 Bell survey needed only 7 hours, on average, to prepare for and complete an initial expedited review, half the time (14 hours) spent on preparing for and completing a full board initial review. IRB meeting time averaged 2 minutes per expedited initial review, compared with 8 minutes per full board initial review.

Expedited reviews are also completed in less elapsed time than are full board reviews: 18 percent of expedited reviews in the 1995 Bell survey were completed in 1 week or less, compared with only 5 percent of full board reviews; 84 percent of expedited reviews were completed in 1 month or less, compared with only 49 percent of full board reviews. (By 3 months' time, over 90 percent of all reviews had been completed; Bell, Whiton, and Connelly, 1998:Figure 33). The savings in elapsed time from expedited review facilitates more timely initiation of research, which can be important for many reasons, including the ability to recruit participants, reduce recall errors in interviews, and meet contractual deadlines. Such savings also conserves on the scarce time of IRB members.

A cautionary note is that IRBs that rarely or never conduct full board reviews—apparently a small group from the available data—may create too casual an atmosphere regarding human research participant protection and undermine trust in the protection system. Similarly, researchers who always seek exemption or expedited review, even when there is a reasonable doubt that the research is less than minimal risk, may undermine the protection system.

IRBs with Very Low Volume

The results presented above on IRB burden and other findings in our report are based primarily on the experience of IRBs that are housed at research institutions that perform significant amounts of research. Almost no information is available on isolated, very-low-volume IRBs, which likely represent a substantial proportion of IRBs but a relatively low proportion of research protocol reviews. One study examined 12 such IRBs associated with community-based hospitals (Office of Inspector General, 1998a). These IRBs conducted a median number of 44 initial reviews per year, with a range of 5 to 124 reviews. The study found that the 12 IRBs experienced workload pressures because of lack of resources. Their members lacked experience with human research participant protection issues and tended to raise fewer questions in review and require fewer modifications to research protocols than IRBs at academic research centers (Office of Inspector General, 1998a, 1998b). These IRBs may be more at risk of insufficient review than of excessive review.

SBES RESEARCH

To this point, we have discussed issues of harm, risk, and benefit, and IRB operations with respect to minimal-risk research from the perspective of the SBES research community without clarifying what we mean by SBES research. To provide context, we conclude this chapter by briefly considering SBES research fields, questions of interest, and commonly used methods. We offer examples of SBES research. Five characteristics of SBES research are important to keep in mind:

1. SBES research is extremely diverse, including classical laboratory experiments, ethnographic research, oral histories, large-scale field experiments, small-scale surveys, large-scale surveys, secondary analysis, other types of methods, and combinations of methods. This diversity can pose challenges for overworked IRBs, particularly in the absence of detailed guidance about how to handle particular situations.[12] Clearly, a "one size fits all" approach is not appropriate, whether the issue is protecting confidentiality, evaluating harms and risks, minimizing risk of harm, or ensuring informed consent.

2. SBES research often does not lead to direct benefits to the participants themselves such as are possible from medical research, but

---

[12] Biomedical research also covers a wide range of topics and methods that pose challenges for the adequacy of IRB review.

much SBES research is minimal risk, especially when an appropriate standard of assessment is applied (e.g., not exaggerating the harm from transitory psychological effects).[13]

3. An important source of risk that must be addressed for many SBES research projects is protecting the confidentiality of individual information. For many research studies in this domain, disclosure risk is the only or the primary form of risk.

4. SBES research uses deception at times as a necessary design element in order to obtain valid results. The Common Rule specifies conditions when deception is appropriate (45 *CFR* 46.116d), which include that the project is judged to be minimal risk and that the research could not produce valid results without the use of deception.

5. SBES research is often embedded in life events (e.g., an ethnographic study of job-seeking behavior in a community, which is carried out over a period of several months or years). Such research usually necessitates that interview protocols and study procedures be modified as the study proceeds. How to accommodate such changes in ways that do not put participants at risk and do not disrupt or delay the progress of the research is a challenge for IRBs. How to maintain informed consent as the study proceeds is also an issue in such research, as well as in longitudinal surveys that follow individuals or families over many years.

SBES Research Fields and Questions

The social, behavioral, and economic sciences encompass a wide range of academic disciplines. While there is no agreement on the precise boundaries, SBES under most definitions includes such disparate fields as cultural anthropology, cognitive science, economics, education research, health services research, history (some fields), political science, psychology, sociology, and survey research.[14]

In terms of questions asked, SBES research is concerned with understanding an ever wider range of attitudes, abilities, behaviors, characteristics, experiences, interactions, moods, perceptions, and statuses of individuals, groups, organizations, and governments. Examples

---

[13]See Chapter 3 for some examples of SBES research conducted several decades ago that were more than minimal risk and violated one or more of the Belmont principles.

[14]See, for example, *International Encyclopedia of the Social and Behavioral Sciences* (Smelser and Baltes, 2001:Table 2), which lists these fields as SBES research or as "related fields." Other related fields listed include archaeology, demography, geography, law, linguistics, and philosophy.

range from political science studies of the determinants and consequences of voting behavior; to anthropological studies of the roles of men and women in family, religious, and civic life; to social psychological studies of how people stereotype others and the kinds of behaviors that are linked to positive or negative stereotypes.

Individual research projects, of course, are not necessarily, nor often, limited to a single discipline or research question. A growing number of research projects are interdisciplinary in nature. Moreover, some SBES research interests increasingly overlap with biomedical research, which has traditionally focused on human physiology, human diseases and their treatment, and human health.[15] For example, a contemporary study on effective regimens for controlling blood sugar would likely involve a multidisciplinary team of biomedical and SBES researchers to examine psychological, social, and cultural factors that might mediate the strictly biochemical effects of different diets or drug dosages. Conversely, for studies of social behaviors, such as the decision to retire, there is growing interest in augmenting traditional social and economic measures with biological health measures for use in explanatory models. Yet another example of merging interests is the collaboration of behavioral psychologists and neurologists to use advanced brain scan techniques to understand the mechanisms by which various stimuli evoke feelings, perceptions, and actions.

SBES Research Methods

SBES research uses a wide variety of research methods. Traditionally, some methods have been more frequently used by some disciplines than others—for example, laboratory experiments in psychology, and observations and unstructured interviews in anthropology. However, most disciplines today encompass multiple methods, and individual research projects often use two or more types of measurement.

Biomedical research also uses a wide range of methods, including surveys and other measurement types that are commonly associated with SBES, but the two domains differ in the frequency of use of particular methods. Both the overlap and differences are evident from the 1995 survey, in which a sample of biomedical and SBES research protocols were categorized by 16 research methods (more than one could be reported per protocol). Thus, while both domains used self-administered questionnaires, 59 percent of SBES protocols did so, compared with only 21 percent of biomedical protocols. Conversely, 25

---

[15]The IOM committee (Institute of Medicine, 2002) implicitly uses this definition of biomedical research in its report, although it is nowhere stated.

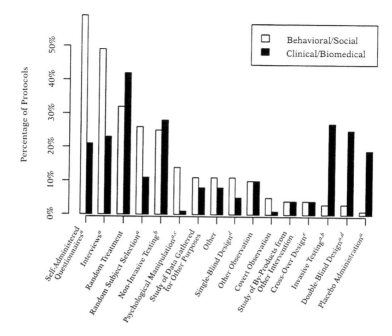

Figure 2-3   SBES and Biomedical Protocols by Type of Method Used

NOTES: Classification by investigators ($n = 632$): behavioral/social research includes social science, behavioral science, educational research, and health services research; clinical/biomedical research includes clinical research, biomedical science, and epidemiology.

[a] Denotes statistically significant difference at the 0.05 level.

[b] Examples of invasive testing include blood sample, biopsy, or spinal tap; examples of noninvasive testing include electrocardiogram or psychological testing.

[c] "Manipulation" means a technique designed to elicit or provoke a response.

[d] In a single-blind design, the subject does not know which treatment is being used but the investigator does. In a double-blind design, neither the subject nor the investigator knows which treatment is being used.

[e] In a cross-over design, treatments are switched between groups during the study.

SOURCE: Bell, Whiton, and Connelly (1998:Figure 8).

percent of biomedical protocols used double-blind experiments compared with only 3 percent of SBES protocols (Bell, Whiton, and Connelly, 1998:16); see Figure 2-3.

Below we briefly describe and provide illustrative examples of some commonly used methods in SBES research—laboratory experiments, field experiments, observations of natural behaviors, unstructured interviews with participants, structured interviews in sample surveys, and analyses of existing data on individuals. For each example, we provide an assessment of the risk of harm and identify other issues of concern for human research participant protection.

Laboratory Experiments

In laboratory experiments an investigator manipulates social or physical conditions in some fashion, and human participants respond to these manipulations (also called treatments or interventions), to which participants are assigned on a random basis. The key purpose of an experiment is to draw inferences about the effect of the intervention on some dependent variable. Participants are not randomly selected; they are recruited in various ways (e.g., newspaper advertisements, students in classrooms) that tend to attract those interested in the purpose of the experiment. See Box 2-1 for two examples—one a typical economic decision-making experiment and the other a social psychology experiment with deception, both of which we judge to be minimal-risk protocols.

Field Experiments

In field experiments an investigator manipulates social or physical conditions in a "real-world" setting to determine the effects on some behavior(s) of human participants. Field experiments are more difficult to carry out than laboratory experiments for at least two reasons: the environment is more difficult to control in the field than in the laboratory, and field experiments are conducted on a larger scale. They may have hundreds or thousands of participants in one or more locations and record measurements at intervals over months or years. The first difficulty makes it desirable to use a control group in addition to one or more treatment groups with participants assigned randomly to a group.[16] The second difficulty usually necessitates a large, multisite team of investigators and elaborate project management. See Box 2-2 for a large-scale welfare policy experiment that is clearly more than minimal risk and a minimal-risk employment discrimination experiment involving deception.

Observations of Natural Behaviors

Observational studies range widely in subject matter. They include the videotaping of interactions among shoppers and store clerks; the

---

[16]Alternatively, a less powerful comparison method may be used (quasi-experiment or natural experiment), such as comparing outcomes for individuals who experienced an environmental change and a comparison group considered to be similar to the treatment group except for experiencing the change.

BOX 2-1
Laboratory Experiment Examples

Economic Decision-Making Experiment The research question is how differing rewards (usually monetary) and rules of behavior affect decision making (e.g., the decision to join a coalition and increase the likelihood of a smaller reward or to stand apart and hope to receive a larger reward that is less certain). A small number of participants (less than, say, 50) are brought together in a laboratory, classroom, or on the Internet. They are given precise, detailed instructions on how they are to interact and how they will be rewarded on the basis of their decisions and the decisions of other participants. They are informed that they may leave at any time. Their decisions are recorded, and they are rewarded accordingly (in private, anonymously, and after the experiment). Reward amounts are small, usually $50 or less. No personal identifiers are kept from the experiment, and no other data (or very limited data, e.g., gender) are collected from participants.

>   *Commentary* This type of experiment is minimal risk—it attempts to replicate commonly encountered decisions (e.g., bargaining over merchandise or votes); the rewards offered are not large enough to be an undue incentive to participate or to cause participants more than momentary dismay (or glee) at the outcome; identifying information is not retained and so the risk that participants' identities could be linked with their decisions is minimal. Such experiments could be exempted from IRB review or reviewed by an expedited procedure. Written consent to participate may not be needed. When students are involved (as is typically the case), care is needed to make sure they understand that deciding not to participate will not affect their grades in the course.

Social Psychology Experiment with Deception The research question is the extent to which people engage in ethnic stereotyping. A small number of participants (less than, say, 50) are brought together in a laboratory or classroom. They are told that the purpose of the experiment is to determine how fast people can associate characteristics (e.g., good, bad) with lists of names (which differ in cues about ethnic origin). Their results are recorded, and they are told at the conclusion of the experiment about its true purpose. No personal identifiers are kept from the experiment, nor are other data obtained about participants except their ethnic origin and perhaps their age and gender.

>   *Commentary* This type of experiment is minimal risk—it attempts to replicate common behavior; there are no incentives to participate; the procedure itself is not stressful; identifying information is not retained; and the inadvertent release of participants' identities and results would cause them only momentary embarrassment at most. The deception invoked covers only the purpose of the experiment, and participants are fully debriefed. Such experiments could be reviewed by an expedited procedure; they could not be exempted given the need to consider the deception involved. Another issue to consider would be fair selection of participants—ideally, a set of experiments would include a range of ethnic groups.

BOX 2-2
Field Experiment Examples

Welfare Policy Experiment The research question of interest is whether welfare recipients are more likely to find a job lasting at least 6 months if they receive training in specific work skills or coaching in job-related behavior skills (e.g., putting together a resume, interviewing for a job, punctuality). In, say, three cities in which coaching is standard practice, the design randomly assigns 2,000 current welfare recipients each to treatment A (skill training) and treatment B (job behavior coaching). Participants are fully informed about the nature of the experiment and that it will not affect their welfare benefits and are promised confidentiality. They are given detailed interviews about their work and welfare history before assignment and reinterviewed at 3-month interviews over a 15-month period (to allow for training and job search time and to observe whether the job lasts at least 6 months). Data from administrative records (time on welfare, benefit amounts, other income) are also obtained. Microdata files containing data for each participant are prepared, stripped of identifiers, and the data processed to minimize the risk that individual participants could be re-identified. The files are deposited with a university archive for secondary analysis.

> **Commentary** This type of experiment is of more than minimal risk, principally because of the extensive amount of data collected, some of which could be sensitive (e.g., if a respondent reports illicit income). The treatment poses little added risk because skill training is (or has been) an expected part of welfare assistance in many jurisdictions. Protecting confidentiality, particularly of the public-use microdata files, is the key concern. Fair selection of participants is also a concern—that is, how the jurisdictions are chosen and whether any are included that offer neither job behavior coaching nor work skill training as standard practice.
>
> If the experiment is evaluating a federal benefits program for a federal program agency, then it is eligible for exemption under the Common Rule (see Box 1-1 in Chapter 1). However, every involved IRB that does not grant an exemption must approve the experiment, or they may delegate review to one institution that has an approved federal-wide assurance (with OHRP's approval of the delegation).

Employment Discrimination Experiment The research question is whether people with felony records experience discrimination in job hiring and whether the effect varies by race. The investigators sample want ads in the Sunday newspapers for factory jobs in a city. They send a pair of candidates to respond to each ad by filling out applications in person at the employer's site. Pairs are assigned to employers randomly. Pairs vary by race (some pairs are two black men, some are two white men); the members of each pair have similar characteristics (age, education, etc.), except that one member of each pair purports to have a felony record. Candidates report if they are called back with a job offer. Employers are not debriefed because it is illegal in the state to discriminate against job applicants who are convicted felons. Data are reported in aggregate form only, and employer identifiers (name, address) are destroyed.

> **Commentary** This experiment is minimal risk because there is no way to link call-back responses to employers and, therefore, no way to embarrass them or subject them to the risk of prosecution. Deception is essential to obtain unbiased responses, and the absence of debriefing about the deception is essential to making the study minimal risk.

visual observation and documentation of conversations within a small group; the audiotaping of conversations of family members; and the visual observation and documentation of neighborhood characteristics. In some observational studies, the researcher's presence and study methods are known to the participants; in others, the researcher attempts to record activities unaffected by knowledge on the part of participants that they are being observed. The purpose of the research is to describe the nature of the activities of those observed; secondarily, the research may argue that the findings would be replicated in other settings. See Box 2-3 for two examples—a minimal-risk study of street-crossing behavior, with no interaction of the researcher and participants, and a more-than-minimal-risk study of the behavior of patrons of a bar, with interaction between the researcher and participants.

Unstructured or Semistructured Interviews with Participants

Unstructured or partially structured interviews are used in a wide range of social and behavioral investigations to elicit an individual's interpretation of events, beliefs, and behavior. Oral histories of public figures or individuals involved in a public event might include unstructured or semistructured interviews. Ethnographers may use informal unstructured interviews during the initial phase of their work.[17] Informal unstructured interviews also may be used throughout a study to build rapport or explore newly emerging topics of interest.

Unstructured interviews have a topical focus but are marked by minimal control over the informants' responses. An investigator working in the area of HIV prevention, for example, may use unstructured interviews with injection drug users to explore beliefs and practices associated with accessing treatment programs. While the general topic has been defined, there is no attempt to follow a predetermined line of inquiry using an interview guide of questions to be asked. In an unstructured interview, the conversation follows the direction taken by the interviewer based on the responses of the interviewee.

In contrast, semistructured interviews involve the use of an interview guide to assist the investigator in systematically studying a partic-

---

[17]Hammersley and Atkinson (1995:248) define ethnographic research to include the following features: "a strong emphasis on exploring the nature of particular social phenomena, rather than setting out to test hypotheses about them; a tendency to work primarily with 'unstructured' data, that is data that have not been coded at the point of data collection in terms of a closed set of analytic categories; investigation of a small number of cases, perhaps just one case, in detail; analysis of data that involves explicit interpretation of the meanings and functions of human actions, the product of which mainly takes the form of verbal descriptions and explanations, with quantification and statistical analysis playing a subordinate role at most."

BOX 2-3
Natural Behavior Observation Examples

Street-Crossing Observation, No Interaction The purpose of the study is to observe street-crossing behavior of pedestrians (e.g., whether they obey walk signs) in relation to such factors as time of day, number of people crossing, and number of cars in the street. The investigator(s) stands at the crossing, makes notes of what occurs (time, number and gender of pedestrians, etc.), and takes still photographs. The investigator makes no effort to interact with pedestrians or to obtain identifiers. Faces are blanked out on the photographs, and they are not published.

   Commentary The project is minimal risk and eligible for exemption. The setting is a public place in which there is no expectation of privacy. The only concern is keeping the photographic material confidential and unpublished (since consent was not obtained).

Observation of Behavior at a Bar by a Regular Customer (Participant Observation) The purpose of the research is to document the social interactions at a neighborhood bar and compare the results with similar studies that have been conducted in other places or other times. The investigator becomes a regular customer, informing the bartender and customers that he or she is doing a book about the bar and will not publish anything in the book about the individual without his or her permission and will not use real names. The investigator conducts research over a 6-month period (typing notes in a palm pilot during restroom breaks and after leaving the bar each evening), writes a book, and carefully reviews applicable sections and statements with each person prior to publication, obtaining signed releases.

   Commentary This project is of more than minimal risk given that sensitive material may be discussed by participants with the investigator and that others may recognize participants despite the use of pseudonyms. A key concern is informed consent and when it must be obtained.

---

ular set of issues. Questions are listed in a specific order and are often followed by leads for exploring the topic in greater detail. Semistructured interviews might be implemented in situations in which the researcher needs to be sure that the same data are collected from all participants. Semistructured interview guides also may be used in focus groups with small numbers of individuals selected on the basis of specific criteria to discuss a particular topic (e.g., a sample of women with known risk factors for breast cancer discussing genetic testing; family members caring for an elderly parent with Alzheimer's disease). A moderator facilitates discussion using a flexible interview guide, and the discussion may continue for 1-2 hours; the conversation is audiotaped and transcribed.

   Data collected from participants using unstructured or semistructured interviews may not be summarized in a statistical fashion. The inference from the data collection is often targeted to a relatively small group (a village, a network of friends, a work group). In these instances, information is valuable because it provides richly elaborated data about the question under investigation. In other cases, researchers may use unstructured and semistructured interviews in addition to quantitative methods to develop an in-depth and robust understanding of a problem. For example, a study of advance care planning in a hospice program might include a survey of a random sample of patients and their families and health providers, along with semistructured interviews with some study participants.

See Box 2-4 for three examples of research using unstructured or semistructured interviews; two of the examples are minimal risk, one of which involves changes in the interview protocol as the study proceeds.

Structured Interviews in Sample Surveys

Much quantitative SBES research involves collection of data in sample surveys, which identify a target population and draw a subset with probability methods to assure that each member of the population has a known, nonzero chance of selection. Once the sample is identified, the researcher cannot substitute other cases without threatening the ability of the research to describe the target population. Participation is sought by having interviewers visit or telephone sample persons or by contacting them by the mail or the Internet. Surveys are designed to describe the attributes of large populations with measurable levels of uncertainty from sampling. See Box 2-5 for two examples—one a minimal-risk telephone survey and the other a complex, more-than-minimal-risk longitudinal survey with linkages to administrative records.

Secondary Analyses of Existing Survey and Records Data

Secondary analyses perform no new data collection. Instead, survey data collected by another researcher are reanalyzed on a topic not previously researched with the data, or records collected for some other purpose are statistically analyzed to study the attributes of a population covered by the record system. Such records may include program agency records, medical records, academic records, and criminal and civil justice records.

Examples of secondary analyses with survey data include studies of labor force behavior with the public-use microdata files from the Census Bureau's Current Population Survey or the University of Michigan's Panel Study of Income Dynamics. These files are preprocessed by the issuing agency to ensure the confidentiality of individuals in the sample.

Examples of secondary analyses with records include studies of income dynamics using Social Security Administration earnings data and studies of occupational mobility using personnel records of a firm. Records data are most often anonymized prior to the analysis. The subjects of the research sometimes cannot be reached by the holder of the record system and do not know of the analysis; other times, they can be reached.

See Box 2-6 for two examples—an analysis of public-use files from a large government survey and a study of school transcript records, both minimal risk.

CONCLUSION

We have discussed a broad array of issues related to the determination of harm, risk, benefit, and minimal risk in SBES research, along with evidence that many IRBs—despite punishing workloads—do not appear to be using the flexibility in the Common Rule regulations to exempt eligible research or to use an expedited procedure to review minimal-risk SBES and biomedical research. Our discussion of types of SBES research illustrates the diversity of the field and the challenge of developing examples to include in guidance. We argue in subsequent chapters for detailed guidance for researchers and IRBs that will improve understanding and encourage the use of the flexibility that is currently possible according to the Common Rule to protect participants in ways that are commensurate with risk. Such guidance will also help researchers design studies that appropriately balance risks and benefits and that incorporate good practices for human participant protection.

BOX 2-4
Unstructured or Semistructured Interview Examples

Epidemiological and Ethnographic Study of Injection Drug Users The goal of this study is to examine the diffusion of benefits associated with injection drug users' participation in needle exchange programs. This multisite project involves more than 500 injection drug users in three cities. Participants are interviewed several times over a 4-year period and agree to have researchers observe them while they are engaged in drug-related activities in order to determine risky behaviors for HIV transmission. Participants also agree to show researchers their drug paraphernalia, including needles used to administer drugs. Both unstructured and semistructured ethnographic interviews are conducted, and a detailed survey is completed. Information on sensitive topics such as drug use history and sexual behavior is obtained. The survey does not include personally identifiable information. Transcripts of interviews do not include names of participants. All data are kept in locked files. Given the vulnerability of the population because of their involvement in illicit activities, two of the sites allow verbal informed consent; however, the third study site requires written informed consent from all study participants.

> **Commentary** The primary risk to participants in this study is the potential breach of confidentiality that could result in stigmatization, physical or emotional harm, or possibly incarceration. This study requires full board approval from an IRB because of the vulnerability of the population and the potential risks involved. However, this example calls attention to the inconsistent application of federal requirements for written informed consent in behavioral studies. The IRBs at the study sites requiring only verbal consent have considerable experience reviewing research proposals addressing social and health behavior of drug users; the IRB requiring written consent has little experience with research on drug users. Given the sensitive nature of the study and the importance of protecting the names of participants, written informed consent should not be necessary to conduct this study. However, investigators should be required to document carefully the procedures used to obtain informed consent and methods for recording the consent discussion.

Case Study of Informed Consent Practices in International Genetic Research A semistructured interview is administered to 20 health professionals in a Nigerian town to explore challenges associated with obtaining informed consent in community-based genetic research being conducted in their area. The study participants are invited to participate in this case study because of their involvement in international scientific investigations. Interviews last approximately 1 1/2 hours. Verbal consent is obtained. Identifying information is removed from the interview transcripts. Transcripts are kept in a locked file.

> **Commentary** This study is minimal risk. Verbal consent is appropriate. The IRB could implement expedited review. The primary risk is the potential for breach of confidentiality regarding sensitive information that may be communicated during the interview. The IRB should require investigators to indicate procedures for obtaining consent, strategies for protecting confidentiality, and, if audiotapes are used during the interview, when they will be destroyed or erased or how they will be protected if they are stored permanently in a research archive.

BOX 2-4 (continued)

Ethnographic Study of Communication about Death and Dying Among Hospice Staff The goal of this study is to explore patterns of communication about death and the dying process among health professionals working in a hospice. The investigator has discussed the study goals and procedures with the hospice administration. After the study has been approved, the staff is informed about the study objectives. Patients and family members are advised of the study. The ethnographer conducts field observations over a 6-month period and conducts unstructured and semistructured interviews with hospice staff. Verbal consent is obtained from all individuals interviewed. If interviews are recorded, consent is recorded at the beginning of the interview. The semistructured interview guide used initially is changed on the basis of responses of the staff to specific questions; some questions are deleted, and new questions added. Field notes are recorded using code names for individuals observed. Interview data and audiotapes are locked.

> **Commentary** This type of study involves minimal risk. The primary risk to participants is the potential for breach of confidentiality, particularly concerning sensitive information regarding communication about death and dying. An IRB would be justified in requiring full board approval because of the sensitive nature of the topic and because patients and families are implicated in the research, even though they will not be interviewed directly. Verbal consent is appropriate when semistructured interviews are conducted. The IRB should ask investigators to outline procedures for obtaining informed consent and strategies for protecting confidentiality, including the disposition of audiotapes if they are used for interviews. The change in the semistructured interview guide should not require full board approval by the IRB. The modified semistructured interview guide should be submitted to the IRB when the study is scheduled for annual review. However, the IRB should be notified if there are any substantive changes in the research design involving major alterations of the methods or the study population (e.g., if the investigator decides to include interviews with patients and family members who are hospice clients after the study has begun).

BOX 2-5
Structured Interview (Sample Survey) Examples

Consumer Telephone Survey To estimate consumer confidence in the economy, a 7-minute telephone interview is conducted of a sample of 1,000 adults in households whose phone numbers are randomly generated by computer software. One adult is selected to report on household plans for purchase of major appliances, savings plans, and opinions about economic prospects for the household and the nation as a whole over the next 6 months. Sample households are repeatedly called until contact is made; interviewers inform respondents that the survey is completely voluntary and address concerns that reluctant respondents may have about participating; no incentive is offered; respondents who initially refuse are called again to seek reconsideration of participating. No names or addresses are collected, and only basic background characteristics are obtained (number of household members, type of household, household income in broad categories). Data are deposited with an archive for secondary use.

> ***Commentary*** This type of survey is minimal risk. It could be exempted from review or reviewed by an expedited procedure. Consent is tacit as is usual in telephone surveys with content that is not stressful and when respondents are informed that they may terminate the interview at any time.

Longitudinal In-Person Health and Retirement Survey To study retirement behavior and health of older adults, a sample of 12,000 adults aged 51-62 in the base year is drawn and interviewed at 2-year intervals (spouses are also interviewed); a new sample is drawn periodically. The interviews are in person; advance letters inform respondents about the survey; the interviews are 1 hour in length each; topics include detailed work history, income, benefits, health status and history, retirement plans and expectations, and other characteristics; incentives are used to promote participation. Data are linked with administrative records, including social security earnings records and descriptions of employer pension and health benefit plans. Some of the data are provided for public use; access to the full microdata requires special arrangements.

> ***Commentary*** This survey is large, complex, and clearly of more than minimal risk because of the sensitivity of the questionnaire content and the risk of breach of confidentiality. Key issues are protecting the confidentiality of the data for secondary use, developing an effective consent process that does not unnecessarily discourage response, and determining an appropriate incentive level to encourage participation.

BOX 2-6
Secondary Analysis Examples

Analysis of Changes in Poverty Levels with the Survey of Income and Program Participation **(SIPP)** The survey is collected by the Census Bureau, which processes the data on public-use files to minimize the risk of re-identification of respondents.

Commentary Even though SIPP obtains highly detailed information on sensitive topics (e.g., detailed sources of income), this is a minimal-risk study that can be exempted from review. The Census Bureau is known to be a leader in confidentiality protection; also, the survey is voluntary, and the Bureau collects the data in an ethical manner. There is no more protection that an IRB can provide for the respondents than the Bureau has already provided in preparing the public-use file.

Analysis of School Transcript Records The purpose of the study is to correlate SAT scores with college grades for recent graduates. The researcher obtains the data, stripped of identifiers, from university registrars; conducts the analysis; and returns the data to the universities.

Commentary This type of analysis is minimal risk given that the researcher has no way of linking student records to individual students. The principal concern is whether the students gave consent for their records to be used for research; another concern is whether students might be identifiable by inference.

— 3 —
Regulatory History

THERE ARE MANY ACCOUNTS that trace the evolution of U.S. policies and regulations for human research participant protection in federally funded research generally and in experimental biomedical research specifically. For example, the report of the Advisory Committee on Human Radiation Experiments (1996) provides a very informative history of regulation and human protection issues in radiation experiments and other biomedical research, covering not only the U.S. Department of Health and Human Services (DHHS), but also the U.S. Departments of Defense, Energy, and Veterans Affairs. But nowhere is there a comprehensive history of human participant protection in social, behavioral, and economic sciences (SBES) research. Problems in such research have generally not received as much media attention as those in clinical trials and other biomedical experiments; also, the diversity of research questions and methods under the umbrella of SBES research is so broad as to make it difficult to construct a thorough-going history (see Chapter 2).

We are not able to make up for the lack of a comprehensive history here. Our summary below focuses on key events in federal policy making and regulation for SBES research from the end of World War II through the present.[1] Our discussion identifies five major periods, the end dates of which mark significant events in the development of regulations, major disputes, and media attention.

The history of federal regulation for human research participant protection shows relatively less emphasis on issues of protecting the confidentiality of information from individual respondents, in contrast to the attention devoted to such issues as the definition of research involving human participants and the elements of informed consent. Im-

---

[1] Key sources that we consulted for this history include Advisory Committee on Human Radiation Experiments (1996:Ch. X); Beauchamp et al. (1982:Ch. 1); Gray (1982); McCarthy (1998); National Bioethics Advisory Commission (2001:App.C). McCarthy (1998) covers attitudes toward and problems with human research participant protection in the United States prior to 1945. See Appendix A for changes in regulatory language from 1974 to the present on such topics as minimal risk, criteria for institutional review board (IRB) review, basic elements of informed consent, exempt research, and others.

portant developments in confidentiality protection, which we consider a major concern for much SBES research, have mainly occurred in federal statistical agencies and SBES data archives. More recently, explicit federal confidentiality regulations have been enacted for research that uses medical records from health care providers and insurers. We review the history of confidentiality protection at the beginning of our discussion of confidentiality issues in Chapter 5.

FROM 1945 TO 1966

The biomedical experiments performed by Nazi researchers during World War II focused attention on the ways in which human participants could be seriously harmed by research. Many unwilling participants suffered permanent injury, shortened life expectancy, psychological trauma, and even death in these experiments (e.g., measuring how long humans could survive in ice water). The second Nuremberg Military Tribunal condemned such research as a crime against humanity. In 1947 the judges proposed a list of ten principles, the "Nuremberg Code," which researchers should be obligated to respect in conducting medical experiments that involve human subjects. The code called for voluntary consent, minimization of harm, and a determination that the research benefits outweighed the risks to participants (see Annas and Grodin, 1992).

Biomedical research exploded in scope and volume after the war. Professional organizations and government agencies began to develop codes of ethics for human research participant protection that reflected the Nuremberg Code. In 1953-1954 the National Institutes of Health (NIH) established an ethics review committee (the Medical Advisory Board) for intramural research at its clinical center and adopted a policy that all human research participants at the center must provide informed consent, although *written* consent was not always required for sick patients. By the mid-1960s many biomedical research organizations had voluntarily established ethics review mechanisms. But instances of unethical research kept coming to light, and public pressure built for explicit government regulation.

The thalidomide tragedy spurred congressional action, as did revelations at congressional hearings about the common practice in which doctors gave patients samples of experimental drugs without the patient knowing the experimental nature of the drugs or consenting to their use. The 1962 Kefauver-Harris amendments to the Food, Drug, and Cosmetic Act required informed consent in testing of experimental drugs, although consent was not required if it was deemed by the physician to be infeasible or not in the patient's best interest. Reports of other questionable research, including a 1963 study in which poor elderly patients at the Jewish Chronic Disease Hospital in New York City were injected with live cancer cells without their consent, led the U.S. Public Health Service (USPHS) to develop a policy on human research participant protection that was issued in 1966.

The 1966 USPHS policy required that every research institution receiving grant dollars from the agency establish a committee to review federally funded research projects for conformance with human participant protection. On December 12, 1966, in response to questions, the U.S. surgeon general issued a "clarification" of the USPHS policy, making explicit that review committees (the forerunners of IRBs) were to review "all investigations that involve human subjects, including investigations in the behavioral and social sciences." With regard to SBES research, he said (Memorandum from the Surgeon General to Heads of Institutions Receiving Public Health Service Grants, December 12, 1966; quoted in Gray, 1982:331):

> There is a large range of social and behavioral research in which no personal risk to the subject is involved. In these circumstances, regardless of whether the investigation is classified as behavioral, social, medical or other, the issues of concern are the fully voluntary nature of the participation of the subject, the maintenance of confidentiality of information obtained from the subject, and the protection of the subject from misuse of the findings. . . .
>
> [SBES research] may in some instances not require the fully informed consent of the subject or even his knowledgeable participation.

The original USPHS policy, as Gray (1982:331) notes, "contained no systematic analysis of the ethical issues at stake in research involving human subjects." Nor did the surgeon general explain why or under what circumstances informed consent in SBES research might not be required.

FROM 1966 TO 1974

The years subsequent to the first USPHS policy guidelines saw continued interest in and attention to human research participant protection issues among regulators, researchers, and members of Congress.

Federal Government Activity

In 1971, in response to requests for clarification and evidence of highly variable implementation of the USPHS policy, the U.S. Department of Health, Education, and Welfare (DHEW)[2] published "The Institutional Guide to DHEW Policy on Protection of Human Subjects." This guide, known as the "Yellow Book," defined "risk" as exposure to the possibility of harm, including physical, psychological, sociological, or other harms, "beyond the application of those established and accepted methods necessary to meet [participants'] needs." (This statement distinguished research activities from activities to provide a service, such as medical treatment.) The Yellow Book explicitly identified kinds of possible harm that could arise in research conducted with such typical SBES methods as interviews, observations, and secondary analysis of existing data, including "varying degrees of discomfort, harassment, invasion of privacy, or... a threat to the subject's dignity through the imposition of demeaning or dehumanizing conditions."

The Yellow Book discussed methods for protecting against disclosure of confidential data. It stated that IRBs should make sure that secondary analysis was "within the scope of the original consent." Such consent could be oral or written, obtained "after the fact following debriefing," or could be "implicit in voluntary participation in an adequately advertised activity." There was no repetition of the surgeon's general statement allowing less than fully informed consent under some circumstances.

Building on the Yellow Book, DHEW issued comprehensive regulations in May 1974 for the protection of human research participants (45 *Code of Federal Regulations* [*CFR*] 46), which stated that the department would not support any such research unless first reviewed and approved by an IRB.[3,4] IRBs were to determine whether human subjects were at risk (defined as the possibility of "physical, psychological, or social injury"); whether risks were outweighed by benefits to the subjects and the importance of the knowledge sought; whether the rights and welfare of subjects were adequately protected (although

---

[2]DHEW was the predecessor agency to DHHS.

[3]The 1974 regulations referred to a "committee" and not to an IRB; the IRB terminology was adopted in a subsequent 1975 amendment to the Public Health Service Act.

[4]Later in 1974 DHEW published regulations for additional protections for pregnant women and fetuses and for prisoners; these regulations were revised in 1978 following the report of the National Commission for the Protection of Human Subjects of Biomedical and Behavioral Research (see below); they became subparts B and C of 45 *CFR* 46, while the basic regulations became subpart A. Additional protections for children in subpart D were first codified in 1983.

there was no specific language about privacy or confidentiality protection); and whether "legally effective informed consent" would be obtained by "adequate and appropriate methods." Informed consent was to be documented in writing. While the regulations included provisions for waiver of written consent, they were difficult to understand and appeared to be very narrowly drawn.

The regulations specified that IRBs were to include at least five members of varying backgrounds, including individuals who were able to "ascertain the acceptability of proposals in terms of institutional commitments and regulations, applicable law, standards of professional conduct and practice, and community attitudes." Gray (1977), among others, argued that this language could give IRBs license to object to research on political rather than ethical grounds.

A government staff member (Tropp, 1979, 1982) later asserted that the pressure of congressional action (see below) resulted in hasty publication of the 1974 regulations, which were formulated primarily by the health components of DHEW, without participation from other agencies that supported SBES research and knew the SBES research community. Further intradepartmental negotiation was planned to produce regulations appropriate for all types of research, but it did not take place.

National Research Act

During this period, public concern about unethical biomedical experiments escalated. Two experiments in particular captured public attention. In the Willowbrook study, from 1956 to 1972, children who were residents of the Willowbrook (Staten Island, New York) State School for the Retarded were injected with a form of hepatitis. Parents, in order to have their children admitted to the only available area of the school, the research wing, had to consent to the study (see Beecher, 1970). In the Tuskegee, Alabama study, begun in 1932, USPHS physicians followed several hundred African-American men who had syphilis. No treatment was given to these men even after the discovery of penicillin. The study was not discontinued until 1973 (see Jones, 1981). The Willowbrook study had been reviewed by an ethics committee, and the Tuskegee study apparently had also had such a review, but neither study was stopped until the media reports and subsequent public reactions.

In 1973 Senator Edward Kennedy (D-MA) held hearings and introduced legislation to establish an independent National Human Experimentation Board to regulate *all* federally funded research with human participants, not just that funded by DHEW. However, this approach did not garner sufficient support, so Senator Kennedy introduced instead what became the National Research Act of 1974. The act endorsed the regulations about to be promulgated by DHEW (see above) and established a National Commission for the Protection of Human Subjects of Biomedical and Behavioral Research. The commission was charged to review the IRB system and to advise DHEW and the Congress about ethical issues in research involving vulnerable populations, such as pregnant women, fetuses, children, prisoners, and institutionalized mentally ill or retarded people. The commission was to identify basic ethical principles for research with human participants and to recommend ways to ensure that research studies followed these principles.

FROM 1974 TO 1981

National Commission

The National Commission for the Protection of Human Subjects of Biomedical and Behavioral Research conducted its work over a 4-year period. In addition to holding hearings and commissioning papers, it contracted with the University of Michigan for an in-depth study of the IRB system at a sample of 61 institutions, involving interviews with several thousand IRB members, researchers, and research participants (Cooke, Tannenbaum, and Gray, 1978; see also Appendix D).

This survey offered three major findings: (1) by comparison with a study for 1969 (Barber et al., 1973), IRBs had become much more active in requiring modifications in proposed research, particularly with regard to informed consent;[5] (2) the selection of participants seemed to distribute risks and benefits fairly evenly among various population groups; and (3) researchers, as well as IRB board members, supported the IRB system overall. The study also revealed two problems: First, not only did virtually all modifications to informed consent procedures deal with consent forms and not the process, but also IRB-modified consent forms were not more complete or easier to read than other forms. Second, the support for the IRB system among SBES researchers was somewhat less strong than among biomedical researchers (see below).

The commission's report on IRBs was issued in 1978; it supported the basic IRB system and made numerous recommendations to streng-

---

[5]The Barber et al. (1973) study, the first to survey IRB operations, interviewed a single individual at each of 300 institutions with a biomedical IRB, 70 percent of which had existed prior to the 1966 USPHS requirement. It found that relatively few IRBs ever required modifications of protocols: 34 percent had never modified or rejected a project.

then and clarify its scope and procedures. Recommendations included: definitions of "research" and "human subject;" an expedited review procedure whereby IRBs could delegate the review of minimal-risk protocols to the chair or a subset of IRB members; an expanded list of types of information to disclose to obtain informed consent; and a clearer specification of the conditions under which written consent could be waived. The commission also recommended that IRBs should: determine that study methods are appropriate (i.e., review protocols for scientific soundness); determine that adequate procedures are in place to protect privacy and protect data confidentiality; not consider possible future harm from research results in making risk-benefit determinations;[6] and provide an opportunity for investigators to respond in person or in writing to IRB decisions. Finally, the commission recommended that DHEW should issue regulations applicable to all research over which it had regulatory authority and that Congress should pass legislation to cover human participants in all research over which the federal government might have regulatory authority.[7]

In 1979 the commission issued the landmark Belmont report, which articulated three basic principles for research studies involving humans: respect for persons, beneficence, and justice. This report quickly became the bedrock on which subsequent work on protecting human research participants was built.

SBES Activity

In both this period and earlier, the SBES research community devoted considerable attention to participant protection issues. A literature review conducted for a 1982 volume on *Ethical Issues in Social Science Research* (Beauchamp et al., 1982) turned up more than 3,500 relevant citations. First and most prominent in the literature was Herbert C. Kelman, who published a collection of early papers in 1968 as *A Time to Speak*. Other leading publications in the years 1967-1980 covered such topics as ethical dilemmas in SBES research, ethics of social experimentation, legal issues in SBES research, politics and SBES research, ethics of social intervention, ethical issues in behavior modification, studying deviance, and confidentiality protection (see Beauchamp et al., 1982:6-7,37-38).

---

[6] Such harm could occur, for example, when results are unfavorable to a population group to which research participants belong.

[7] Congress never acted on this recommendation; legislation is pending at present that would extend federal protections to all research involving human participants (see "Developments Since 1991," below).

BOX 3-1
Examples of Ethically Troubling SBES Research from the 1970s

Milgram's (1974) "Obedience to Authority"

Summary Stanley Milgram, a social psychologist, carried out an experiment in the 1960s in which volunteers were recruited to be "teachers" who were asked to administer an electric shock of increasing intensity to a "learner" for each mistake made during the experiment. The fictitious story given to these "teachers" was that the experiment was exploring effects of punishment (for incorrect responses) on learning behavior. The "teacher" was not aware that the "learner" in the study was an actor who merely indicated discomfort as the "teacher" increased the (fake) electric shocks. When the "teacher" asked whether increased shocks should be given, he or she was verbally encouraged to continue. Sixty percent of the "teachers" obeyed orders to punish the "learner" to the very end of the 450-volt scale; no subject stopped before reaching 300 volts. Many subjects were distressed by what they were asked to do. Subjects were debriefed at the end of the experiment and offered counseling if they wanted to examine their own behavior and feelings. Milgram later carried out variations of this same experiment.

Commentary Milgram's experiments aroused intense controversy about the conditions under which it is appropriate to use deception, particularly when the consequence is intense (although apparently transitory) psychological stress and even when subjects are informed about the deception at the conclusion of the experiment. Milgram subsequently resurveyed his original subjects, most of whom said they supported the experimentation because it provided important information about how ordinary people could be induced to behave in reprehensible ways in response to a recognized authority, but this finding did not stop the controversy.

Humphreys' (1975) "Tearoom Trade"

Summary Laud Humphreys, a graduate student in sociology, conducted a study for his dissertation of homosexual behavior among men of high social standing in a large Midwestern city. Pretending to be gay and a lookout for the police, he observed men entering a public rest room in a city park, confirmed that they engaged in anonymous homosexual acts, recorded their license tag numbers, obtained their names from a contact in the Bureau of Motor Vehicles, and confirmed their identity and social status. A year later he disguised himself and interviewed them in their homes about their personal lives under the pretext of conducting a social health survey. When his dissertation was published, many of the men and their families were able to recognize themselves from the details that were reported about them. Humphreys' research had the perhaps beneficial result that police stopped raiding "tea rooms" when they learned that many of the men frequenting them were not people they wanted to arrest.

Commentary Had the Common Rule been in effect at the time of "Tearoom Trade," his research would not likely have been approved. Given the stigma attached to homosexuality, the study would not be categorized as minimal risk, and, therefore, deception could not be used (see Box A-9 in Appendix A on the criteria for waiver of informed consent).

BOX 3-1 (continued)

Whether norms have changed to the extent that such a study might be viewed as minimal risk today is not clear; however, informed consent would certainly be required before study results could be published, and other changes in study procedures would undoubtedly be required before it could be approved.

Zimbardo's (1971) Stanford Prison Experiment

Summary In summer 1971 psychologist Philip G. Zimbardo studied the reactions of male volunteers (Stanford University students) to a mock prison environment in which some of them were assigned roles as prisoners, others as guards. The experiment elicited such strong reactions from the participants, including abuse of some "prisoners" by some "guards," that Zimbardo halted the 2-week study after only 6 days (see Faden and Beauchamp, 1986:178-179; http:/www.prisonexp.org [4/10/03]).

Commentary This study did not involve deception. Whether such a study would be approved today is unclear. Evidence from Milgram, Zimbardo, and others that strong negative reactions are not uncommon would likely make IRBs reluctant to approve further research of this type. However, an investigator could propose steps to minimize risk, such as planning prompt interventions of the type that were implemented by Zimbardo in the prison experiment, that could make the study acceptable, and it would have the potential to achieve important results for understanding social interactions in stressful situations.

Social scientists and other researchers also engaged in vigorous debates about the ethics of particular research projects, including the Milgram "obedience to authority" experiments, Humphreys' study of the "tearoom trade," and Zimbardo's study of prison behavior (see Box 3-1). None of these projects caused long-lasting physical harm to any participant, but there was concern that they may have caused strong short-term psychological effects and perhaps even long-lasting psychological trauma. However, for none of them, to our knowledge, has there been any documented lasting harm to participants.

As a consequence of these debates and the growing concern with human participant protection in SBES research, both the American Anthropological Association and the American Sociological Association adopted new codes of ethics in 1971. The American Psychological Association, which had first adopted a code of ethics for clinical work and research in 1953, sponsored extensive research and discussion on ethical issues specific to psychological research and adopted a revised ethics code in 1973 (Faden and Beauchamp, 1986; see also Chapter 4).

With regard to the 1974-1978 work of the national commission, there is no available evidence that concerns about SBES research

---

BOX 3-2
SBES Concerns in the 1970s

Problems of 1974 IRB Regulations for SBES Research

- IRBs' insistence on *written* consent when it was not needed and would damage the research.
- Infringements of academic freedom resulting from IRBs' application of risk-benefit criteria.
- The general lack of fit between the regulations and the ethical issues that arise in some types of research (e.g., ethnography).
- The failure of the regulations to address explicitly certain issues (e.g., deception, privacy, confidentiality of data, and the use of research for teaching purposes) and certain types of research (e.g., fieldwork or evaluation research).
- The conflict that can develop between written consent requirements and researchers' responsibility to assure confidentiality of the identity of subjects.
- The lack of clear definition of the applicability of the regulations (e.g., just what *is* a human subject).
- The harm to research that can result from application of consent requirements to certain research, such as observational studies in public places and studies based on existing records.
- The use of IRBs to protect vested interests (e.g., the alleged tendency of IRBs in some medical schools and hospitals to discourage social research within their walls).
- Some IRBs' lack of reasonable flexibility in interpreting regulatory language.
- Some IRBs' lack of qualifications to review certain types of studies.
- The application of review requirements to research not funded by the federal government.
- The lack of appeal procedures.

SOURCE: Verbatim quotes from Gray (1982:354) of testimony provided to the National Commission for the Protection of Human Subjects of Biomedical and Behavioral Research.

---

played a part in its formation (the "behavioral research" part of its title was not explicitly defined in its charter). As specified by Congress, the commission membership included six people who were not researchers and five who were: three physicians and two psychologists were chosen to fill the researcher slots. Although the commission did not represent the breadth of SBES expertise (or biomedical expertise, for that matter), it did seek input from SBES researchers. Most of their concerns, according to Gray (1982:335), "pertained to harm done to research interests while providing no benefit to the interests of subjects" (see Box 3-2). There were differing viewpoints: a paper commissioned from a sociologist who served as chair of an IRB that primarily reviewed SBES research argued that current regulations did not impede research, provided they were interpreted reasonably (Barber, 1979); a paper commissioned from another sociologist argued that existing regulations did not properly address issues of informed consent and confidentiality in SBES research (Reiss, 1979).

Data from the University of Michigan survey conducted for the commission revealed that almost all researchers believed the IRB system had helped protect human participants (at least to some extent), although higher percentages of SBES researchers than of biomedical researchers were critical of the system in some respects. For example, excluding IRB board members, 54 percent of SBES researchers agreed that the IRB system had impeded the progress of research at their institution (at least to some extent), compared with 43 percent of biomedical researchers (Gray, 1982:Table 16.2).[8]

1979 Proposed Revision of Regulations

In response to the national commission's recommendations, the NIH Office for Protection from Research Risks (OPRR) drafted a revision of 45 *CFR* 46, which was published for comment in the *Federal Register* in August 1979.[9] The proposed revision raised a firestorm in the SBES research community. The major problem appears to have been that the regulations proposed to extend the IRB system to all research involving human participants at institutions that received DHEW funds, regardless of whether the particular protocol was funded by DHEW, and to studies in all fields that sought generalizable knowledge by using methods that collected information by which living *or dead* individuals could be identified. (See Boxes A-1 and A-3 in Appendix A, which trace changes in language on applicability of the regulations and the definition of human subject, respectively.) This broad scope could have subjected standard historical, document-based research, interviews with public officials, observation of behavior at public events, and other such studies to IRB review, including student and investigator studies funded by the research institution itself or by another federal agency and not DHHS. Furthermore, IRB review was explicitly to consider the appropriateness of the research methods for

---

[8] Researchers who were IRB members were much less likely to agree that the system impeded research (30 percent of SBES IRB board members did so, as did 26 percent of biomedical IRB board members).

[9] OPRR was established in the office of the NIH director in the early 1970s; it superceded the Institutional Relations Branch of the Division of Research Grants, which formerly had responsibility for oversight of human research participant protection (McCarthy, 1998:313).

the objectives of the research and the field of study (see Box A-6 in Appendix A).

The proposed revised regulations included provisions for IRBs to exempt certain types of research from review and to review other kinds of research by an expedited procedure (see Boxes A-4 and A-5 in Appendix A). The intent of these provisions was to exempt or expedite much SBES research, such as that conducted by surveys, public observation, and study of documents. The exemption provisions were restricted to research in which it was not possible to identify the human subjects (Alternative A) or that presented no or only minimal potential for invasion of privacy (Alternative B). Similarly, the expedited review procedures were restricted to minimal-risk projects in specified categories, and the proposed definition of minimal risk set a high standard—namely, that risks of harm be no greater than those experienced in the daily lives of *healthy* individuals (see Box A-11 in Appendix A). SBES researchers argued that all of these features of the proposed regulations subjected too many projects to full-scale review by an IRB when such review was not necessary or useful or even (in some arguments) constitutional.

On the positive side, the proposed regulations responded to some SBES concerns that had been expressed in testimony before the national commission. Specifically, provisions were added to require IRB review to consider the adequacy of proposed data confidentiality protections, to enable IRBs to modify some or all of the elements of informed consent, and to enable IRBs to waive the requirement for written documentation of consent (see Boxes A-6, A-9, and A-10 in Appendix A). These generally well-received changes (which remained in the revised regulations) were overshadowed, however, by the negative reactions to provisions that SBES researchers viewed as inimical to good research.

Patullo (1982:373) characterizes the ensuing period until January 1981, when 45 *CFR* 46, subpart A, was revised, as follows:

> For eighteen months, hundreds of individuals, some principally concerned with maintaining a healthy research enterprise and some alarmed at basic incursions upon accustomed freedoms, devoted thousands of hours to persuading those who write the regulations that single-minded concentration on precluding every possibility of harm to subjects had produced a proposal that threatened much greater social harm than it might possibly prevent.[10]

---

[10]Patullo was among those who argued against the 1979 proposed regulations on the basis that they threatened research and also among those who argued that the regula-

The head of OPRR, which drafted the regulations, gave his view of this tumultuous period as follows (McCarthy, 1984:8-9):

> Threats and epithets were hurled at us from many sides. These comments can be summarized as follows: 'Social and behavioral research is essentially harmless to individuals and benign to society. If you do not accept this view... we will organize ourselves to see that you and your misguided staff are drawn and quartered.' The charges were led by Ithiel de Sola Pool [1979, 1980], who insisted that... our four pages of fine print in the *Federal Register* were about to lay waste to the First Amendment of the Constitution. ...
>
> Friendly champions of social and behavioral sciences showed us how to back away from our unpopular positions while continuing to offer what we felt were reasonable protections for the dignity and rights of subjects involved in social and behavioral research—to say nothing of saving the face and the jobs of OPRR staff.
>
> What were the indignities to subjects that we felt needed attention? The Wichita jury bugging case, the tearoom trade research, the decision of a Georgia court concerning Medicaid co-payment experiments—and our own unpleasant memories of Psychology 101 and Sociology 102, when we felt we had better humor our professors... so we 'volunteered' as research subjects rather than risk grade discrimination!
>
> We discovered that we could write exemptions for broad categories of social and behavioral research—categories in which subjects' behavior seemed to us little different than the commerce of daily life. It has been estimated that up to 80% of social and behavioral research funded by our Department is now exempt. For the rest, we thought it not unreasonable to concern Institutional Review Boards with matters of privacy and confidentiality, and with efforts to protect unsuspecting and vulnerable subjects.

The final regulations, issued in January 1981 by DHHS (which replaced DHEW in May 1980) responded both to the recommendations of the national commission (see above) and the concerns raised by SBES researchers. With regard to the latter, the regulations somewhat narrowed the definition of a human subject and provided for certain

tions were developed without evidence that SBES research caused physical harm (see Oakes, 2002:7).

kinds of human subjects research to be exempt from IRB review. A "human subject" was defined as "a *living* individual about whom an investigator (whether professional or student) conducting research obtains (1) data through intervention or interaction with the individual, or (2) identifiable *private* information" [italics added]. This definition excluded some SBES research from IRB review, such as observational studies in which private information was not obtained and studies of information from past public records. In addition, the regulations provided that four categories of human subjects research could be exempted by an IRB from review (see Box A-4 in Appendix A):

- research in educational settings on new or established instructional strategies and techniques, curricula, or classroom management methods;

- research involving educational tests, if identifiers cannot be linked to the subjects;

- survey and interview research and observational research, unless identifiers are being collected, disclosure could be damaging to the participant, and the data pertain to sensitive subject behavior; and

- studies using existing data, documents, records, and the like, if these materials are publicly available or if the data will not be recorded in a manner that would allow linkage with individuals.

Finally, backing off from the 1979 proposal, the regulations applied only to research sponsored by DHHS, although language was included indicating DHHS's concern that the interests of all human research participants, regardless of funding source, be protected at institutions that received DHHS funding.[11] Also backing off from the 1979 proposal, the regulations excluded a requirement for IRB review of the appropriateness of the proposed research design and methods, requiring only that IRBs determine that risks to subjects are minimized "by using procedures which are consistent with sound research design."

One observer concluded that "most social research is now exempt from the regulations either because it is specifically exempted or because it is not federally supported, [however], the extent to which IRBs

---

[11] In backing off from this position, which had been a recommendation of the national commission, DHHS responded to concerns of the SBES community, a growing opposition to federal regulation in general, and a specific conclusion of the 1980 President's Commission for the Study of Ethical Problems in Medicine and Biomedical and Behavioral Research that the National Research Act of 1974 could not be used to justify extension of DHHS regulations to non-DHHS-funded research.

will actually continue to review such research remains to be seen"
(Gray, 1982:344).

FROM 1981 TO 1991

With the adoption of the 1981 regulations, a period of relative calm
ensued. A handful of studies of IRB operations were carried out. One
study, which surveyed 341 IRB chairs soon after the passage of the
regulations, found that almost all IRBs at that time had decided *not*
to use the option to exempt protocols from review that fell under one
of the four eligible categories of educational, social, and behavioral
research (Grundner, 1983).

A somewhat later study examined how IRBs were functioning from
the perspective of political scientists; it queried 115 chairs of political
science departments that offered Ph.D. programs and chairs of IRBs at
those institutions (Cleary, 1987; see also Appendix D). Generally, most
chairs were supportive of the IRB system, reporting that most political
science protocols (88 percent) cleared without change. Problems identified included uncertainty and lack of information regarding informed
consent and confidentiality protection; confusion at some institutions
as to whether unfunded student research was covered; and variability
in IRB procedures with regard to exemptions and informed consent.

In 1981 the President's Commission for the Study of Ethical Problems in Medicine and Biomedical and Behavioral Research (which met
from 1980 to 1983) recommended that a Common Rule be developed
that would apply to all federally supported research involving human
participants. In response, in 1982 the President's Science Adviser, Office of Science and Technology Policy, appointed an interagency committee to develop a common framework on the basis of the DHHS regulations. The committee published a proposed common policy in 1986.
Its work culminated on June 18, 1991, when 45 *CFR* 46, subpart A,
was adopted in the regulations of 15 departments and agencies and by
the Central Intelligence Agency by legislation (see Box 1-1 in Chapter
1). The Food and Drug Administration also modified its regulations to
agree in large part with 45 *CFR* 46, subpart A.

In addition to making the Common Rule regulations applicable
to research conducted, supported, or regulated by any of the federal
agency signatories, the 1991 regulations incorporated several other
changes from the 1981 DHHS version (see Appendix A). First, the definition of research was expanded to include research development, testing, and evaluation. Second, two categories of research that could be
exempted were added: research and demonstration projects involving

public benefit or service programs and taste and food quality evaluation and consumer acceptance studies. Third, the requirements for exempting survey and observational research were modified so that IRBs could more readily decide *not* to exempt as much research as before (see Box A-4 in Appendix A). The list of types of research that were eligible for expedited review, provided they were deemed to be of minimal risk, was not changed at that time, but it was later amended in a *Federal Register* notice in 1998: many more categories of SBES research were added to the list than had been included in the original (1981) list (see Box A-5 in Appendix A).

DEVELOPMENTS SINCE 1991

Raising the Alarm

Even as the federal regulatory system for protection of human research participants was strengthened by the widespread adoption of the Common Rule, a spate of reports in the mid- to late 1990s concluded that IRBs were less and less able to meet their responsibilities. These reports dealt almost exclusively with concerns arising from biomedical, particularly clinical, research.

The first such report emerged from the Advisory Committee on Human Radiation Experiments, convened in 1994 to follow up media reports of federally sponsored radiation research with human participants conducted between 1944 and 1974 that violated ethical norms.[12] The committee also investigated the current state of human participant protection by reviewing the regulatory system, examining a sample of recent research proposals, and interviewing past, current, and prospective research participants. While concluding that "significant advances" had occurred in human participant protection since the 1940s and 1950s (Advisory Committee on Human Radiation Experiments,1996:510), the committee found "evidence of serious deficiencies in some parts of the current system." These included substantial variability in the performance of IRBs and inattention to problems of informed consent for people with diminished decision-making capability, as well as confusion among participants as to whether they were involved in experimentation or treatment.

The U.S. General Accounting Office (GAO) in 1996 found that the ability of IRBs to operate effectively was impaired by heavy IRB workloads, lack of expertise on IRBs to review complex research, failure

---

[12] Such experiments included injecting hospitalized patients with plutonium, likely without their knowledge or consent, and intentional release of radiation into the environment without public notice.

of IRBs to exercise continuing review over research projects, and lack of adequate facilities and support for IRB operations. The GAO also found that OPRR lacked time and funding to conduct site visits to review IRB operations, even though the value of such visits in identifying IRB compliance problems was known.[13]

Beginning in 1998 the DHHS Office of Inspector General issued a series of reports warning that IRBs were being overwhelmed by the complexity and volume of research they had to review. The first report (Office of Inspector General, 1998b:ii-iii) concluded that "the effectiveness of IRBs is in jeopardy" because "they face major changes in the research environment," "they review too much, too quickly, with too little expertise," "they conduct minimal continuing review of approved research," "they face conflicts that threaten their independence," "they provide little training for investigators and board members," and "neither IRBs nor HHS devote much attention to evaluating IRB effectiveness." Furthermore, the report claimed that IRBs lacked adequate resources to do their job and were being pushed by regulations to focus on paperwork requirements more than on basic ethical issues.

The National Bioethics Advisory Commission, established by presidential appointment in 1995, concluded in its final report (2001:4-8) that several factors underlay these gloomy assessments of the IRB system. They included:

- a large increase between 1985 and 1996 in the volume of research resulting from a two-fold increase in federally funded research and a three-fold increase in industry-funded research (particularly by pharmaceutical firms);

- a growing propensity for academic medical centers to develop ties with industry for research funding;

- an increase in industry-sponsored research conducted through nonacademic organizations;

- the emergence of stand-alone IRBs that operate on a fee-for-service basis;

- a growing number of clinical studies conducted at multiple, sometimes hundreds, of sites;

---

[13]Between 1990 and 1995, OPRR conducted only 15 site visits during the course of over 200 compliance investigations triggered by charges of misconduct. No site visits were conducted randomly, nor were any visits conducted as part of assurance negotiations, when OPRR first reviewed an institution's written procedures for conformity with the Common Rule (U.S. General Accounting Office, 1996:19).

- and new technologies, such as research on genetic links to diseases and new gene therapies for medical treatment, and changes in public attitudes[14] that pose new ethical challenges.

All of these factors overloaded IRBs with no offsetting increases in the resources available for their work.

Federal Response

In response to these reports and to media attention on tragic events in biomedical studies (see below), the federal government stepped up its regulatory efforts. In June 2000, the Office for Human Research Protections (OHRP) was established in the Office of the Secretary of DHHS, taking over responsibilities from OPRR and acquiring new responsibilities (see http://ohrp.osophs.dhhs.gov [4/10/03]; 67 *Federal Register* 10217, March 6, 2002). OHRP was given a broad mandate, not only to monitor the operations of IRBs that review DHHS-funded research, but also to provide guidance on human research participant protection for the federal and nonfederal sectors, develop educational programs, and exercise leadership for human participant protection in cooperation with other federal agencies. A standing committee of outside experts, the National Human Research Protections Advisory Committee (NHRPAC), was established to provide continuing advice to OHRP.

Prior to the establishment of OHRP, OPRR had become somewhat more active in enforcement: from one on-site inspection in 1997, OPRR conducted ten on-site inspections in 1998-2000 (all at medical research centers). Several of these reviews resulted in temporary suspension of all or some research at the visited institutions (Office of Inspector General, 2000:2).[15] OHRP emphasized working cooperatively with research institutions and associations to develop training programs for IRBs and investigators and to improve IRB operations, but it, too, on occasion took drastic action. In July 2001 OHRP suspended nearly all research at Johns Hopkins University when a healthy young woman participating in a study about asthma medications died from an over-

---

[14]Examples are research that is conducted with the assistance of community organizations and initiatives by some very ill patient groups to obtain greater access to clinical trials, seeing participation as a benefit, not a burden.

[15]In May 1999, OPRR suspended all research at the Duke University Medical Center for 5 days and required several corrective actions, including re-review of DHHS-funded research approved by the IRB, establishment of a second IRB, and development of educational programs for IRB staff and members and researchers (National Bioethics Advisory Commission, 2001:56).

dose of an asthma-inducing substance (see Keiger and De Pasquale, 2002).[16]

In these and other widely publicized incidents, such as the death of Jesse Gelsinger in 1999 in a University of Pennsylvania gene transfer experiment, it appeared that the researchers failed in one or more respects to follow the protocol approved by the IRB. Nevertheless, the events increased calls for reform of the IRB system, such as recommendations by the National Bioethics Advisory Commission (2001:xi-xxi) for legislation to establish human participant protection for all publicly and privately funded research and a single independent federal office to lead and coordinate the oversight system; a single set of federal regulations and guidance; accreditation of IRBs; and certification of IRB members.

In fall 2001 DHHS commissioned the Institute of Medicine to undertake a thorough review of the IRB system, including proposed schemes for voluntary accreditation of IRBs by nonprofit organizations (Institute of Medicine, 2001, 2002). Congressional committees held hearings, and Representatives DeGette (D-CO) and Greenwood (R-PA) introduced a bill in May 2002 that would extend federal human participant protection to all federally funded research and require the harmonization of the Common Rule and Food and Drug Administration regulations. Senator Kennedy (D-MA) introduced a bill in October 2002 that would expand protections to all research involving human participants. It would also strengthen OHRP, providing for a 6-year term for the director.

SBES Involvement

During this same period, researchers involved in SBES studies increasingly raised concerns that IRBs were overreacting to the increased scrutiny of their operations by putting minimal-risk studies through as time-consuming a review as clinical trials and other higher risk studies and by rigidly adhering to the main provisions of the regulations rather than taking advantage of their options for flexibility. Some argued (although less strongly than in earlier debates about IRBs and SBES research) that the IRB system infringed on academic freedom to conduct research (see, e.g., Shea, 2000). Many argued that IRBs were unnecessarily impeding research and sometimes increasing the risk to participants by their bureaucratic stance—for example, by requiring written consent even when such consent could endanger participants

---

[16] In all, work on 2,400 protocols was halted, some for 5 days and some for longer periods. Research at Johns Hopkins' Homewood campus, which operated under a different federal assurance, was not affected.

by linking them to the study (see, e.g., American Association of University Professors, 2001; Brainard, 2001; Moreno, 2001; Shea, 2000). In 1998, at the behest of the SBES community, additional categories of minimal-risk SBES research were added to the list of research categories that IRBs could review under expedited procedures (see Box A-5, Appendix A).

SBES researchers also fought to become involved in key groups that were part of the debate and that often initially excluded SBES disciplines. In January 2001, NHRPAC (the advisory body to OHRP) was expanded from 11 to 17 members in order to include SBES researchers. In June 2001, NHRPAC established an ad hoc Behavioral and Social Science Working Group, which is currently operating as a separate entity with private foundation funding. In September 2002, DHHS declined to renew NHRPAC's 2-year charter. A new Secretary's Advisory Committee on Human Research Protections was chartered in October 2002 with 11 members. The membership of the committee was announced in January 2003; it includes only two SBES researchers, both psychologists.[17]

The National Science Foundation (NSF), a principal funder of SBES research, established an ad hoc Social, Behavioral, and Economic Subcommittee for Human Subjects. The subcommittee developed a guide for SBES researchers, which was put up on the NSF Policy Office website as "Frequently Asked Questions and Vignettes: Interpreting the Common Rule for the Protection of Human Subjects for Behavioral and Social Science Research" (National Science Foundation, 2002). A new Association for the Accreditation of Human Research Protection Programs (AAHRPP) added 3 seats for SBES researchers to its 21-member board and involved SBES researchers in pilot tests of accreditation procedures at several dozen research institutions.[18] (See Appendix B for information about groups active in and resources for human research participant protection.)

---

[17] See http://www.hhs.gov/news/press/2003press/20030113a.html [2/23/03]. In December 2002 the presidents of the National Academy of Sciences, the National Academy of Engineering, and the Institute of Medicine wrote a letter to DHHS Secretary Thompson stating that "it is critical for this important body to represent all of the scientific disciplines engaged in research with human participants—that is, the biological, clinical, social, behavioral, and economic sciences."

[18] AAHRPP was established in May 2001 as the national accrediting arm of Public Responsibility in Medicine and Research. It is developing a voluntary, peer-driven human research accreditation program, using a site visit model; see Appendix B.

CONCLUSION

This brief history makes clear that the concerns of the SBES research community about the IRB system are not new. In fact, the lists of concerns voiced in 1974 (see Box 3-2) are remarkably similar to those being voiced today. Moreover, both today and three decades ago, some researchers believe that the regulations need to be changed, while others maintain that the regulations have sufficient flexibility and the real concern is with their interpretation.

We conclude that the current focus for improvement of participant protection and facilitation of research in SBES should be on the development of guidance and other means to encourage IRBs to avail themselves appropriately of the flexibility in the regulations rather than on further changes in the regulations themselves. A primary reason for our conclusion is the past successes of the SBES research community with respect to regulatory change. Thus, in the late 1970s, SBES researchers successfully participated in a process that resulted in the current provisions for exemption and expedited review. In the late 1990s, SBES researchers influenced the process that resulted in an expanded list of types of research that, if minimal risk, could be reviewed with an expedited procedure. The regulations also provide for flexibility in procedures for obtaining and documenting informed consent.

As we discuss in subsequent chapters, OHRP and other actors in the protection system, such as accreditation organizations and professional associations, must assert leadership in providing guidance that can help IRBs use the flexibility in the Common Rule to carry out their responsibilities more effectively. To assure the development of appropriate guidance, SBES researchers must actively continue to seek formal and informal channels for input to OHRP and other relevant agencies and organizations. The SBES research community must also actively seek to develop a knowledge base that can inform OHRP, IRBs, and researchers about appropriate procedures for informed consent, balancing of risks and benefits, and other aspects of ethically responsible research.

— 4 —
Enhancing Informed Consent

I NFORMED VOLUNTARY CONSENT is one of the bedrock principles of ethical research with human participants. The 1947 Nuremberg Code affirmed the principle of voluntary consent to participate in research, and U.S. and international policies and regulations from the beginning incorporated *informed* consent as a key element of ethical research (see Chapter 3). Following the Belmont report (National Commission for the Protection of Human Subjects of Biomedical and Behavioral Research, 1979), which articulated the principle of respect for persons, and as that principle gained in importance, federal regulations elaborated the requirements for informed consent and its documentation.

By "informed consent," we mean that a person's decision to participate in research is made by the individual (or by his or her legally authorized representative), without pressure to hurry the decision, without coercion or undue influence from the investigator to participate, and with relevant information about the research that is provided in understandable language (see Faden and Beauchamp, 1986). For minimal-risk research, a process that allows consent to be truly informed chiefly respects individual autonomy (voluntariness). For more-than-minimal-risk research, such a process is critical not only for autonomy, but also to allow the individual to make reasoned judgments related to potential harms and benefits of participation.

In the Common Rule provisions for human participant protection, two of the seven criteria for institutional review board (IRB) approval of research (45 *CFR* 46.111—see Box 1-1 in Chapter 1) are that informed consent will be sought and that it will be appropriately documented, usually by a participant's signing a written consent form in advance of participation. Other sections of the regulations specify elements of informed consent and special conditions when consent elements may be waived (see Boxes 4-1 and 4-2); and required documentation of informed consent, including when advance written consent may be waived (see Box 4-3).

BOX 4-1
Basic Elements of Informed Consent

(a) **Basic elements of informed consent**. Except as provided in paragraph (c) or (d) of this section, in seeking informed consent the following information shall be provided to each subject:

  (1) a statement that the study involves research, an explanation of the purposes of the research and the expected duration of the subject's participation, a description of the procedures to be followed, and identification of any procedures which are experimental;

  (2) a description of any reasonably foreseeable risks or discomforts to the subject;

  (3) a description of any benefits to the subject or to others which may reasonably be expected from the research;

  (4) a disclosure of appropriate alternative procedures or courses of treatment, if any, that might be advantageous to the subject;

  (5) a statement describing the extent, if any, to which confidentiality of records identifying the subject will be maintained;

  (6) for research involving more than minimal risk, an explanation as to whether any compensation and an explanation as to whether any medical treatments are available if injury occurs and, if so, what they consist of, or where further information may be obtained;

  (7) an explanation of whom to contact for answers to pertinent questions about the research and research subjects' rights, and whom to contact in the event of a research-related injury to the subject; and

  (8) a statement that participation is voluntary, refusal to participate will involve no penalty or loss of benefits to which the subject is otherwise entitled, and the subject may discontinue participation at any time without penalty or loss of benefits to which the subject is otherwise entitled.

SOURCE: Verbatim quotes from 45 CFR 46.116 (a) (boldface added for ease of reference).

Over time, changes to federal regulations have expanded and modified the provisions on informed consent (see Boxes A-7 through A-10 in Appendix A). New items were added to the list of basic information that must be provided to participants; a list was added of informational items that can be provided to participants when appropriate; language was added about circumstances under which required elements of informed consent can be modified or waived; and language was clarified about the circumstances under which documentation of informed consent can be waived.

In this chapter we begin with a review of the available evidence on the ways in which IRBs interpret the informed consent provisions of the Common Rule and the problems, often unintentional, that a narrow interpretation can cause for research that uses such methods as surveys, interviews, observations, laboratory experiments, and analyses of existing data. Such methods are commonly used in the social, behavioral, and economic sciences (SBES) and are also frequently used in biomedical research. We also review the available evidence on how consent procedures may affect comprehension of risks and benefits and participation in research and recommend a research program to improve consent procedures and forms for different types of SBES research and populations studied.

The remaining sections of the chapter present our analysis and recommendations on useful guidance the Office for Human Research Protections (OHRP) should develop for IRBs (which can also aid researchers) on applying the Common Rule provisions on informed consent. We consider four topics: informed consent issues for certain vulnerable and special populations (e.g., language minorities); issues regarding consent of third parties (i.e., people who are not being directly interviewed or observed in a study); when it is appropriate and good practice to waive written signed advance consent; and when one or more elements of informed consent may be omitted from the consent process and documentation.

We stress that informed consent is best thought of as a *process*, whereby an investigator interacts with potential participants to inform them about study goals, harms, risks, and benefits, and other pertinent information about the project (see Institute of Medicine, 2002:Ch. 4). The process should be designed to respect individuals' rights and to allow them to make informed choices about the risks and obligations they are willing to accept to participate in a research project. Depending on the nature of the research, the consent process may require explaining and obtaining consent for the research in advance, or at the time when data collection actually begins, or repeatedly throughout the project to be sure the participant understands the steps in the research and their implications. In contrast, *documentation* of consent, while important, should be viewed as secondary to the consent process. Documentation should be tailored to facilitate and not impede or confuse the actual process of obtaining informed consent, which may often— but not always—include obtaining a signed written consent form.

We urge OHRP to begin immediately to work with relevant professional associations, IRBs, investigators, and representatives of research participants to develop detailed guidance, with examples for types of research and populations studied, on informed consent processes and to add to and modify the guidance as needed in the future (see discussion in Chapter 7). Without authoritative guidance on such topics as when it is appropriate to waive written consent and in a cli-

---

BOX 4-2
Additional Elements of Informed Consent and Provisions for Waiver or Alteration

(b) **Additional elements of informed consent.** When appropriate, one or more of the following elements of information shall also be provided to each subject:

  (1) a statement that the particular treatment or procedure may involve risks to the subject (or to the embryo or fetus, if the subject is or may become pregnant) which are currently unforeseeable;

  (2) anticipated circumstances under which the subject's participation may be terminated by the investigator without regard to the subject's consent;

  (3) any additional costs to the subject that may result from participation in the research;

  (4) the consequences of a subject's decision to withdraw from the research and procedures for orderly termination of participation by the subject;

  (5) a statement that significant new findings developed during the course of the research which may relate to the subject's willingness to continue participation will be provided to the subject; and

  (6) the approximate number of subjects involved in the study.

(c) **An IRB may approve a consent procedure which does not include, or which alters, some or all of the elements of informed consent set forth above, or waive the requirement to obtain informed consent** provided the IRB finds and documents that: (1) the research or demonstration project is to be conducted by or subject to the approval of state or local government officials and is designed to study, evaluate, or otherwise examine: (i) public benefit service programs; (ii) procedures for obtaining benefits or services under these programs; (iii) possible changes in or alternatives to those programs or procedures; or (iv) possible changes in methods or levels of payment for benefits or services under those programs; and (2) the research could not practicably be carried out without the waiver or alteration.

(d) **An IRB may approve a consent procedure which does not include, or which alters, some or all of the elements of informed consent set forth above, or waive the requirements to obtain informed consent** provided the IRB finds and documents that:

  (1) The research involves no more than minimal risk to the subjects;

  (2) The waiver or alteration will not adversely affect the rights and welfare of the subjects;

  (3) The research could not practicably be carried out without the waiver or alteration; and

  (4) Whenever appropriate, the subjects will be provided with additional pertinent information after participation.

SOURCE: Verbatim quotes from 45 CFR 46.116 (b), (c), (d) (boldface added for ease of reference).

BOX 4-3
Documentation of Consent and Waiver Conditions

(a) **Except as provided in paragraph (c) of this section, informed consent shall be documented by the use of a written consent form** approved by the IRB and signed by the subject or the subject's legally authorized representative. A copy shall be given to the person signing the form.

(b) **Except as provided in paragraph (c) of this section, the consent form may be either of the following:**

(1) A written consent document that embodies the elements of informed consent required by 46.116. This form may be read to the subject or the subject's legally authorized representative, but in any event, the investigator shall give either the subject or the representative adequate opportunity to read it before it is signed; or

(2) A short form written consent document stating that the elements of informed consent required by 46.116 have been presented orally to the subject or the subject's legally authorized representative. When this method is used, there shall be a witness to the oral presentation. Also, the IRB shall approve a written summary of what is to be said to the subject or the representative. Only the short form itself is to be signed by the subject or the representative. However, the witness shall sign both the short form and a copy of the summary, and the person actually obtaining consent shall sign a copy of the summary. A copy of the summary shall be given to the subject or the representative, in addition to a copy of the short form.

(c) **An IRB may waive the requirement for the investigator to obtain a signed consent form for some or all subjects if it finds either:**

(1) That the only record linking the subject and the research would be the consent document and the principal risk would be potential harm resulting from a breach of confidentiality. Each subject will be asked whether he or she wants there to be documentation linking the subject with the research, and the subject's wishes will govern; or

(2) That the research presents no more than minimal risk of harm to subjects and involves no procedures for which written consent is normally required outside of the research context.

In cases in which the documentation requirement is waived, the IRB may require the investigator to provide subjects with a written statement regarding the research.

SOURCE: Verbatim quotes from 45 CFR 46.117 (boldface added for ease of reference).

mate of fear in which IRBs may be blamed for any harm that occurs to research participants, IRBs will have no incentive to use the flexibility in the Common Rule to develop the most beneficial consent procedures for particular research protocols (see discussion in Chapter 6). OHRP guidance, informed by research on effective informed consent procedures, will also help researchers improve practice on this vitally important element of human research participant protection.[1]

IRB FOCUS ON INFORMED CONSENT

Data from the available studies of IRBs indicate that issues of informed consent account for a large share of the time and resources that IRBs devote to reviewing research protocols and that investigators, in turn, devote to revising their protocols in order to secure IRB approval. The 1975 Michigan survey carried out for the National Commission for the Protection of Human Subjects of Biomedical and Behavioral Research found that 40 percent of research protocols reviewed by the sampled IRBs between July 1974 and June 1975 were modified as a result of the IRB review process: the most common modification, required for 24 percent of protocols, concerned informed consent (Gray, Cooke, and Tannenbaum, 1978:1096-1097). Almost all such modifications involved the content of consent forms; less than 1 percent involved the consent process (e.g., timing or setting of consent, who obtained consent).

The 1995 Bell survey (Bell, Whiton, and Connelly, 1998) reported data on protocol modification in terms of IRBs, not protocols; those data show clearly that, by 1995, IRBs were more likely to modify protocols and to require changes to informed consent documents than they had been in 1975. Thus, 34 percent of IRBs in 1995 required modifications to *every* protocol in their workload (Bell, Whiton, and Connelly, 1998:61), compared with only 14 percent of IRBs that required modification of every protocol in their workload in 1975 (Gray, Cooke, and Tannenbaum, 1978:1096).[2]

---

[1]OHRP provides access to an *IRB Guidebook* on its website at http://ohrp.osophs. dhhs.gov/irb/irb_guidebook.htm [4/10/03]. The guidebook was originally developed in 1981 for the President's Commission for the Study of Ethical Problems in Medicine and Biomedical and Behavioral Research and updated in 1993; it has useful information and guidance on such topics as informed consent, but the guidance is quite general for the most part.

[2]The 1995 Bell survey data are not consistent as reported. As reported, 73 percent of IRBs disapproved 75 percent or more of protocols as submitted, another 10 percent disapproved between 50 and 75 percent, and 6 percent disapproved fewer than 50 percent.

As of 1995, IRB chairs were most likely to report deficiencies in protocols involving the language of consent forms—60 percent said such a deficiency occurred often; the next highest percentages were 13 percent of chairs reporting that cost information was often omitted from consent forms, and 11 percent of chairs reporting that consent forms often omitted or understated risks of participation. Investigator reports of most commonly required protocol modifications confirmed the IRB focus on consent forms: 78 percent of investigators said they were required to change the consent form, and 21 percent said they were required to change consent procedures.[3]

Applying Common Rule Consent Provisions

Few quantitative data are available on how IRBs apply the informed consent provisions of the Common Rule. The 1975 Michigan survey does report that informed consent was obtained in almost 90 percent of projects, usually in writing. For the remaining projects, investigators said that the return of questionnaires implied consent, that the project used only routine procedures, or that it used existing records gathered for other purposes.[4] The 1975 survey also reported that some information was withheld from participants in 15 percent of studies, usually to eliminate sources of bias or because of the belief that the participant would not understand the information. Omitted information usually pertained to a specific medication or treatment (e.g., in a double-blind study) or the purpose of specific procedures. In 2 percent of studies, participants were given false information, usually concerning the purpose of procedures and usually to avoid bias in the results.

Although quantitative data are lacking, anecdotal evidence suggests that IRB practices with regard to informed consent—particularly documentation of consent—are one of the sorest points for researchers who are engaged in minimal-risk research. For example, Sieber, Plattner, and Rubin (2002) conclude:

These numbers add up to 89 percent, not 100 percent. However, this discrepancy does not invalidate a conclusion that IRBs were more likely in 1995 than in 1975 to require modifications to protocols.

[3]The large percentage of protocols for which modifications were required to consent forms suggests that IRBs could provide clearer guidance to researchers on appropriate content and language (see Chapter 7).

[4]Gray, Cooke, and Tannenbaum (1978:1101) conjecture that "in some cases investigators who reported that they did not obtain informed consent may have meant only that they did not use a consent form. The confusion of the substance of consent with its documentation is not an uncommon error and may have negative implications for informed consent."

In the current regulatory climate, many IRBs treat all social and behavioral research as if it were very risky. They interpret the Common Rule as literally as possible, ignoring any cultural or procedural inappropriateness this may entail, and generating an extensive paper trail to prove that they have done what they construe the Common Rule to require. . . .

Some results of this environment of fear include: (a) a self-defeating quest for entirely risk-free research in a world where nothing is entirely risk free, (b) long delays in approving protocols, and (c) extremely bureaucratic interpretations of the requirement of informed consent. These three problems are intertwined. The focus on very minor or unlikely risks has resulted in lengthy negotiations between IRBs and investigators, and overly detailed, insultingly paternalistic informed consent procedures.

Sieber, Plattner, and Rubin (2002) cite the following examples provided to them by researchers of inappropriate IRB actions with respect to documentation of consent:

- requiring participants who are members of a preliterate society to read and sign a consent form;

- requiring written consent of members of cultural groups who consider it insulting to sign an agreement, as if their word were not to be trusted;

- requiring written consent in situations when oral consent or implicit consent is adequate (e.g., answering a survey question) and requiring written consent only serves to discourage otherwise willing participants;

- requiring written consent in advance of a mail survey of individuals who had previously agreed to be contacted for such research; and

- requiring written consent of students who are asked by their instructor to make judgments and discuss their reasons as a pedagogical exercise in which no data are collected and no research is conducted.

Is the IRB Focus on Informed Consent Effective?

Given the intense focus of IRBs on informed consent, particularly consent forms, three key questions are relevant: First and foremost, does it increase the protection that is afforded to human research participants? Second, does a focus on consent result in improved consent forms? Third, does it help or hinder research in other ways? The available data suggest that consent forms are not understood by many participants and that IRB-required revisions to consent forms do not improve their readability or understandability. There is no evidence on whether IRB attention to consent forms improves the level of protection for participants, but there is some evidence that requiring advance signed written consent inhibits participation for some people who would otherwise not hesitate to join a minimal-risk research study.

Readability of Consent Forms

The 1975 Michigan survey devoted considerable attention to evaluating the consent forms that were used by the research projects included in the study. Only 18 percent of the forms were complete or nearly complete when measured on an index consisting of six elements required by the 1974 regulations: the purpose of the research, the procedures involved, the risks, the benefits, a statement that participants can withdraw from the research, and an invitation to ask questions. The purpose was omitted from 23 percent of the forms, and risk was omitted from 30 percent (however, 70 percent of this 30 percent were said by the investigators to involve a very low probability of minor harm to participants).

The Michigan survey also developed a "reading ease score" in which the "standard" was the language used in *Time* magazine. Only 6 percent of the consent forms analyzed were as easy or easier to read than *Time*; 17 percent were "fairly difficult" (at the level of *Atlantic Monthly*); 56 percent were "difficult" (scholarly); and 21 percent were "very difficult" (scientific or professional). This distribution was similar across types of institutions—universities, medical schools, hospitals, and other—although university consent forms were somewhat easier to read than others (13% were at the level of *Time* or below). There was almost no correlation between the completeness of consent forms and their readability.[5]

Turning to the effects of IRB review, there was no significant difference in average completeness or readability scores between con-

---

[5] Of course, the appropriate readability level relates to the population to be studied. Informed consent materials for research involving university faculty would not need the same degree of readability as research involving the general population. Other issues of readability include type size, which may need to be larger for older people, and appropriate translation for people who are not literate in English (see "Informed Consent for Special Populations" below).

sent forms as proposed to IRBs and consent forms as approved by IRBs. Moreover, looking only at consent forms that were modified in response to an IRB, there was also no significant difference in completeness or readability scores before and after IRB review for these forms either. Finally, the less readable or less complete forms were no more likely than the more complete or more readable forms to be singled out for change by IRBs.

Whether the effects of IRB review of consent forms are still as negligible as found by the 1975 Michigan survey is not known. One study of all approved consent forms at two universities in 1988 and 1991 suggests that IRBs were no more effective in that period than 15 years earlier (Goldstein et al., 1996). This study found that the average reading score of 284 consent forms was 12.2, or roughly a 12th grade reading level, which is above the average reading level of the general population (about 10th grade). Fewer than 10 percent of all consent forms were written at a 10th grade level or below. Readability scores were not related to whether the consent form was revised at the behest of the IRB, the year of the study, or the university. However, survey studies had more readable consent forms than clinical drug trials, and having a higher number of female than male IRB members related to more readable forms.

Research on alternative consent forms in biomedical research has found that readability is not the only issue. For example, a study in which participants had low levels of literacy were given a standard cancer trial consent form (written at a college level) and a simplified form (written at a 7th grade reading level). The participants overwhelmingly preferred the simpler form and found it easier to read. However, their comprehension was low regardless of which form was used (Davis et al., 1998).

It is clear that even after years of research on informed consent, largely covering written consent forms for biomedical research and treatment (see bibliographies in Sugarman, McCrory, and Hubal, 1998; Sugarman et al., 1999), there are still no agreed-upon procedures that are demonstrated to provide adequate, comprehensible information to prospective research participants. Furthermore, it is clear that, on average, IRB efforts to improve participant protection by focusing on consent forms have had relatively little effect and, as such, have diverted scarce time and resources of IRB members and investigators alike, particularly when empirical information is lacking about effective ways to improve consent forms and procedures.

Effects of Consent Requirements on Participation

There is a small literature that helps illuminate the effects of informed consent requirements on willingness to participate in research. For research that is of minimal risk, understanding how information presented in certain ways could discourage participation of individuals who would otherwise be willing to volunteer is important for developing consent procedures that inform participants but do not unnecessarily degrade the research by adversely affecting the size or composition of the sample. Such understanding is also helpful for determining when it may be appropriate and good practice to waive elements of informed consent. For research that is of more than minimal risk to participants, understanding the effects of different informed consent procedures is critical for developing the most effective means for clearly communicating the risks and benefits of participation.

Data exist on participation in surveys and laboratory experiments, primarily in the SBES domain, as well as the desires and concerns of participants regarding the consent process. Singer (1993) reviews the literature published before 1993;[6] Singer's (2003) is the first relevant empirical study to be reported since then.

With regard to the detail provided in survey introductions, Singer (1993:363) sums up early practice as follows:

> Prior to the spate of studies on informed consent procedures in the late 1970s (National Research Council, 1979; Reamer, 1979; Singer, 1978a, 1978b, 1984; Singer and Frankel, 1982), conventional survey wisdom had advocated keeping the introduction *short*, so as not to lose the respondent's interest or attention; and some evidence from experiments with mail questionnaires had suggested that a *general* explanation of purpose was preferable to a more detailed one, which might antagonize some respondents (Blumberg, Fuller, and Hare, 1974). At the same time, some investigators supported fuller disclosure of research purposes to respondents. ... Some support for the efficacy of fuller disclosure came from a study by Hauck and Cox (1974), in which refusals were reduced after respondents had been given a more nearly complete and accurate description of the study's purpose.

Singer's review (1993:365) of studies conducted from 1978 through the early 1990s of the effects on survey participation of providing dif-

---

[6] Work on informed consent issues in survey research was initially stimulated by enactment of the 1974 regulations for protection of human research participants (see Chapter 3) and by enactment of the Privacy Act of 1974 (Dalenius, 1983). Declining response rates also played a role (Steeh, 1981).

ferent amounts of detail about the content and purpose of the survey concluded that, "Within the limits tested, information about content has no perceptible effect on response rates or quality." However, from results of self-administered questionnaires given to respondents at the conclusion of their participation in the main survey, she concluded that "respondents who are given more information about sensitive content are more likely to report, in retrospect, that they expected the questions and that they were not upset or embarrassed by them; they also show less measured anxiety." (The latter finding is from a laboratory experiment: Holliman et al., 1986.)

With regard to the form of consent (written, oral, tacit), Singer (1993:369) found some evidence that "so-called 'passive' consent methods capture the intentions of most potential respondents, and that 'active' methods exclude some who in fact are willing to participate, but not to sign their name." For example, Singer (1978a) reported that 7 percent of respondents in an experiment who were asked to sign a consent form refused to do so: they were not unwilling to participate; they were unwilling to sign the form. These findings were replicated by Trice (1987). Similarly, Ellickson and Hawes (1989) and Moberg and Piper (1990) found that most people who failed to return a signed consent form in the mail did not intend to refuse participation (based on a subsequent telephone query).

In September 2000, Singer (2003) undertook a small study of survey respondents (n = 275) to ascertain their understanding of informed consent, including perceptions about the kind and degree of risk involved, and to ascertain the relationship between respondents' attitudes and their behavior. Questions were added to 1 month's sample of interviews in the University of Michigan Survey of Consumer Attitudes (SCA). Respondents to the SCA were read hypothetical introductions to two surveys fielded at the University of Michigan—the National Survey of Family Growth, conducted under contract to the National Center for Health Statistics, which deals with sexual behavior, pregnancy, childbearing, schooling, work, and medical care; and the Health and Retirement Survey, conducted with a grant from the National Institute on Aging, which deals with personal finances, changes in health, and health care needs. The introductions were similar to those actually used but modified to make statements about risks and benefits as comparable as possible. Half the introductions requested a signature on a consent form and the other half did not.

The results indicated (as in earlier research, e.g., Singer, 1978a) that respondents do not understand or remember everything in the introduction but, given their perceptions, they act rationally. In the Singer (2003) research, respondents' perceptions of risks, benefits, and risk-benefit ratios related significantly to their expressed willingness to participate in the survey described to them. This research also replicated earlier findings with regard to the effects of requiring written signed consent. At least 13 percent of the respondents in the Singer (2003) study said they would be willing to participate but would not be willing to sign a consent form. Singer (2003:21) concludes:

> This finding is crucial to the argument that IRB's should permit researchers to modify the way consent is documented, especially in a study where the risk is minimal. The request for a signature does not appear to protect respondents' rights; on the contrary, it may subvert their expressed desire for participation. And it reduces the generalizability of survey findings, which depend on accurate measurement of all the designated members of the sample.

RESEARCH TO IMPROVE CONSENT PROCEDURES

Recommendation 4.1: Social, behavioral, and economic science researchers should conduct research on procedures for obtaining and documenting informed consent that will facilitate comprehension of benefits, harms, and risks of harm, confidentiality protection, and other key features of research protocols for different types of SBES research and populations studied.

There are many challenges to developing effective consent procedures that truly support a voluntary, informed decision to participate (or not) in research and that do not place inappropriate barriers in the way of participation. Despite several decades of research on consent forms and practice (mostly in biomedical research), and despite intensive IRB attention to documentation of consent, there appears to have been little progress in devising more effective forms and procedures for achieving informed consent. Research on consent practices in SBES research is limited.

We believe it is incumbent on SBES researchers to seek support for sustained research efforts on informed consent that would inform good research practice and also enable OHRP to develop useful guidance for IRBs. The National Institutes of Health recently had a program to study ethical issues in research, such as informed consent procedures and issues. This program should be continued and expanded. Support for research on informed consent for different types of SBES research methods, settings, and populations studied (including such populations as immigrants, refugees, language minorities, and people engaged in illegal or deviant behavior) is also needed.

Such research should include experiments with different kinds of consent processes. For example, a comparison could be performed among three alternative consent procedures: having an investigator engage a prospective participant in conversation about the research without requiring the individual's signature on a form;[7] having an individual read and sign a standard consent form; and having an individual read and sign a consent form designed on the basis of cognitive research and testing for maximum readability and ease of understanding. Measured outcomes could include the degree of comprehension of the material, the number and type of questions the prospective participants ask about the research, individuals' willingness to participate, and (for participants) the extent of satisfaction with the decision to participate in the research. Similar outcomes could be measured in studies that compared how often and at what stages participants are reminded of the nature of the research and given opportunities to ask questions. Studies could also be conducted on effective ways to train investigators in the design and implementation of appropriate consent procedures.

INFORMED CONSENT FOR SPECIAL POPULATIONS

Recommendation 4.2:   The Office for Human Research Protections should develop detailed guidance for IRBs and researchers on appropriate consent procedures for different types of populations—including language minorities and such vulnerable groups as undocumented immigrants— studied in social, behavioral, and economic sciences research.

The process of informed consent emphasizes the elements of disclosure, competence, comprehension, and, finally, a voluntary decision to consent or refuse to participate. Too often ignored are the underlying assumptions embedded in the notion of voluntary informed consent— assumptions about language and the meanings attached to words and concepts and assumptions about relationships and the social position of individuals in families, institutions and communities. Obtaining consent that is both informed and voluntary may be challenging for researchers working in international settings and with populations who are vulnerable because of their immigrant or refugee status or because

---

[7]Documentation of consent in this instance could consist of having the investigator document the discussion and attest that participation was the individual's own decision.

they are involved in illegal activities or deviant behavior, as we discuss below.

Informed consent for SBES research involving infants, young children, and adolescents raises special issues as well. Because we considered informed consent and other ethical practices for research with human participants in the context of the Common Rule (subpart A of 45 *CFR* 46), which applies to the general population, our report does not address special concerns for children (or other people with diminished capacity for informed consent, such as mentally retarded people). We note that subpart D of 45 *CFR* 46 provides added protection for children of three kinds. First, if children are involved, then survey, interview, or observational research that would be exempt from IRB review if it only involved adults may not be exempted. (An exception is observational research in which the researcher does not participate in the activities being observed.) Other kinds of eligible research (e.g., research involving educational tests) may be exempted even if the research is conducted with children. Second, research with children requires both the assent of the child and the informed consent of the child's parents or guardian for participation. Third, research of more than minimal risk for children may only be conducted if the research satisfies a series of additional criteria (see 45 *CFR* 46.405, 406, 407). For in-depth discussions of informed consent and other ethical practices in SBES research with children, see Fisher and Tryon (1990); Grodin and Glantz (1994); Sieber (1992:Ch.10); and Stanley and Sieber (1991).

Consent Issues for Certain Vulnerable Populations

Social science and economic studies of illegal behavior (e.g., drug abuse) or highly sensitive topics (e.g., alcoholism, sexual abuse, or domestic violence) require special attention to two key issues. First, it is vitally important that measures are in place to protect the privacy and confidentiality of research participants (see discussion in Chapter 5) and that these measures are appropriately addressed in the information that is provided to prospective participants in seeking their consent. Second, it is important that care is taken to minimize the potential for coercion, by which we mean a threat to take away a privilege or freedom or to discriminate against someone who refuses to participate. Successful negotiation of informed consent means that individuals are able to participate freely and voluntarily. The perception on the part of a potential participant that refusal to be in a study could result in negative social, economic, or legal consequences diminishes the possibility of voluntary consent. The obverse of coercion is also important to consider—namely, an inappropriate incentive, by which we mean a promise to provide a special privilege.[8] Such a promise (e.g., to intervene on the person's behalf in a legal action) could induce someone to participate in a risky study for which the person might otherwise decline participation.

Issues associated with the protection of confidentiality and concerns about coercion are also important considerations for investigators working with refugee or immigrant populations or conducting research in international settings. The issue of protecting confidentiality remains key, particularly in studies involving illegal immigrants because of the implications for incarceration or deportation. The potential for coercion may exist if individuals do not understand what it means to be involved in a research project—because of language barriers or diverse cultural beliefs about the nature of research and the benefits associated with it—or because threats are used in the recruitment process (e.g., a threat to hinder a refugee's application for permanent status). Conversely, a promise of special treatment (e.g., an offer to speed up a refugee's application for permanent status) could compromise voluntary, informed consent.

Another important factor for investigators to consider when working with refugee or immigrant populations or conducting international research is the emphasis placed on personal autonomy in the Western paradigm of informed consent. In addition to language barriers and different beliefs concerning the nature of scientific research, effective communication may be thwarted when participants and researchers have diverse opinions about who has the authority to decide whether or not someone can participate in a study. In the United States, individuals are expected to make decisions for themselves if they are mentally competent adults, and, given parental permission, minors have the right to assent to or decline research participation. However, in many places throughout the world, decisions are not necessarily made by the individual, but instead by family members or community representatives. Beliefs about personhood, individual autonomy, and decisional capacity are embedded in the social context of family ties and community obligations. In some cultural settings, religious or tribal leaders, the household head, or a person's extended family may play a significant role in major decisions.

The challenges of meeting U.S. requirements for written documentation of informed consent in studies with immigrant populations or in international settings may be particularly problematic when study par-

---

[8]See discussion of financial incentives, which are ordinarily appropriate in SBES research, in "Survey Research" below.

ticipants are reluctant to formalize a document with their signature or thumbprint because of previous experiences that resulted in their victimization, including the loss of personal property or land when "legal" documents were used against them. Verbal consent would be appropriate for minimal-risk research with these populations.

Language, Translation, and the Use of Interpreters

Misunderstandings and miscommunication about social science and economic research are more likely to occur when investigators and potential participants speak different languages and when informed consent documents must be translated. The situation is exacerbated when there are no equivalent expressions for particular concepts or when the concept of informed consent is unfamiliar. The language used to describe the purpose of a study and its associated risks and benefits may be confusing and, in some cases, intimidating for the individual being asked to participate in a research project. Two dimensions of language are important to consider in obtaining informed consent: first, the choice of words, that is, the specific terms that will be used in the consent discussion; and, second, the use of language to express concepts related to the research itself and such concepts related to informed consent as "voluntary participation" and "confidentiality" (Marshall, 2001).

Investigators working with populations who speak a different language (e.g., immigrant or refugee populations in their own country or abroad) depend on accurate and meaningful translations of informed consent documents. A process of back-translation is recommended when translation of consent forms is required: researchers translate the consent form from one language to another, then it is translated back to the original language by someone else, preferably by someone who is not associated with the study. Investigators then evaluate the form to determine that it has communicated accurately and effectively the meaning of the research. Such translation and back-translation may need to be carried out by translators who live in the locality and participate in the culture of the study population in order to ensure that the translated consent form is appropriate.

In some cases, an interpreter may be employed to obtain consent from participants in research. The use of an interpreter may significantly reduce linguistic barriers, but potential problems remain (Barnes et al., 1998; Kaufert and O'Neil, 1990; Kaufert and Putsch, 1997; Marshall, 1992a, 1992b; Marshall et al., 1998; Putsch, 1985). The investigator must rely on the translator to communicate the research objectives correctly. Translators are often portrayed as straightforward interpreters of information exchanged between researchers and potential research participants. This perspective, however, underestimates the complexities of interpretation in which the translator must negotiate not only language, but also cultural and contextual factors (Carrillo, Green, and Betancourt, 1999; Kaufert and O'Neil, 1990; Kaufert and Putsch, 1995; Marshall, 1992b; Marshall and Koenig, 1996; Putsch, 1985). Challenges associated with interpretation include: the inability to translate equivalent expressions across languages; paraphrasing language that results in omissions or erroneous substitutions of terms; different levels of comprehension among participants in the interaction; and the influence of diverse cultural beliefs and values about research among participants. Moreover, if family members or friends act as interpreters, there may be a tendency for them to exaggerate, camouflage, or minimize information (Putsch, 1985).

Finally, cultural norms governing the structure and content of interactions between researchers and those invited to participate in studies are vitally important to effective communication (Barnes et al., 1998; Kaufert and O'Neil, 1990; Kaufert and Putsch, 1997; Koenig and Gates-Williams, 1995; Marshall and Koenig, 1996). Beliefs and expectations regarding what is considered to be "appropriate" discussion in interactions vary considerably across cultures and are affected by social factors that reinforce differences in the relative power experienced and expressed by the individuals involved in the interaction. The topics discussed, the timing of the conversation, and who participates in the conversation influence profoundly the process of informed consent. For example, in many cultural environments, women are subordinate to their husbands, fathers, or male heads of households. In these situations, particularly if the person obtaining consent is a man, women may not believe it is right to question the investigator about the study. Perhaps the most serious negative consequence of this form of deferral to "authority" is that some individuals may not believe they have the right to refuse to participate.

Language and communication are powerful tools that affect what actually occurs in the informed consent discussion. The process of informed consent is always situated in a cultural context, reinforced and constrained by the dynamics of social and political power.

THIRD-PARTY CONSENT

Recommendation 4.3: The Office for Human Research Protections should develop detailed guidance for IRBs and researchers, including specific examples, on when it is and is not necessary to obtain consent from third parties about whom participants are asked to provide information.

Much SBES research involves collecting data from and solely about individual participants—for example, surveys of individual attitudes or knowledge of public affairs and laboratory experiments in which responses to specific stimuli are recorded. Quite often such studies will also collect data on background characteristics of the individual, such as age, race, education, and personal income.

Not infrequently, however, SBES research may ask participants to provide data about other people. Sometimes the data about others are needed for context and to develop a richer explanatory model— for example, information on characteristics of a person's family, such as number of members, relationships, ages, educational levels, and family income in studies of work-leisure tradeoffs or voting behavior. Sometimes the data about others are the main object of research—for example, studies of the effects of parental childrearing practices (as perceived by the participant) on the participant's adult experiences.

Traditionally, household surveys, including major federal government surveys (e.g., the Current Population Survey, the Consumer Expenditure Survey), have asked one household member to provide information for all members of the household about such topics as employment, marital status, expenditures, and income. For more accurate reporting, some surveys (e.g., the Survey of Income and Program Participation) strive to obtain self-reports for each household member but accept proxy responses for household members who are away or otherwise cannot be contacted in the allotted time. Oral histories and ethnographic studies typically ask respondents for considerably detailed information about other people in their families, communities, or other social groups.

Recently, concerns about third-party consent for studies that ask respondents for information about others (third parties) have become more prominent in discussions of ethical research. In particular, discussion has centered on a study of twins who were asked for detailed information about their parents, including sexual characteristics and medical history, which resulted in a public complaint from a parent of a participant that the research violated his privacy and his consent should have been obtained (see Botkin, 2001).

Determining when it is necessary to obtain third-party consent is not straightforward. Such determination requires careful attention to the level of risk of the research, the setting in which it is conducted, the nature of the data to be obtained on others, and the characteristics of the study population. No hard-and-fast rules can or should apply.

Several illustrative situations and our conclusions regarding the appropriateness of third-party consent follow.[9]

- **Respondent Attributes** When information about third parties represents an attribute of the respondent, such as a perception or attitude, then third-party consent usually is not relevant and should not be sought. For example, studies of participants' *perceptions* of parental childrearing practices or *perceptions* of supervisor ethics do not require third-party consent because data being collected are attributes of the respondent. Whether the data are accurate or not regarding the third party is not relevant; the study is examining the reasons for or effects of participants' perceptions of others.

- **Third-Party Anonymity** Some studies may inquire about one or more of a group of people who interacted with the participant, such as the participant's teachers, employers, or sex partners. When the questions are framed in such a manner that the third parties remain anonymous to the researcher as well as to others, then consent is rarely if ever necessary for their protection. For example, a series of questions might ask about a participant's "first supervisor" without identifying the individual in any way. Of course, if the "first supervisor" is also the person's current supervisor, then the issue of third-party consent is less easily dismissed. If there are adequate measures for confidentiality protection to guard against disclosure of third parties as well as respondents, then it is likely that third-party consent is not needed. The nature of the questions being asked will also be a factor in a decision on consent.

- **Authorized Proxy Response** When the desired respondent cannot participate because of disability or other incapacity and an authorized representative of the individual is present, then consent can be obtained from that representative as provided in the Common Rule (see Box 4-3).

- **Household Proxy** When surveys or participant observation research pertain to an entire family or household, the issue is whether consent must be obtained from every member for one member to respond for the family or household as a whole. (Similar issues are discussed below for ethnographic research in group

---

[9] See also the statement of the National Human Research Protections Advisory Committee adopted at its January 28-29, 2002, meeting; available at http://ohrp.osophs.dhhs.gov/nhrpac/documents [4/10/03].

settings.) When the research is minimal risk and does not require written consent—for example, most mail and telephone surveys—then we do not believe it is necessary to obtain third-party consent for one respondent to provide answers for other family members or the household as a whole. Research deemed to be minimal risk should not be harmful to the respondent or any third party, and there is a reasonable expectation when a family member agrees to respond to such research that he or she is trusted to do so for other family members. (Of course, the designated respondent may provide less accurate information about other members than the members would themselves, but that is an issue of data quality and not of informed consent.)

For more-than-minimal-risk surveys or observational research on families or households, it may often be the case that self-reporting should be the goal and that informed consent should be obtained from each reporter. In some situations, it may be necessary to interview some family members at different times from others in order to protect privacy, as well as to enhance data quality. For example, a study of spousal relations and perceptions might conduct separate as well as joint interviews of the two spouses. The consent material provided in such a study should clearly inform each spouse of the intent to interview each about the other.

WAIVING WRITTEN CONSENT

Recommendation 4.4: The Office for Human Research Protections should develop detailed guidance for IRBs and researchers—with clear examples for a variety of social, behavioral, and economic sciences research methods and study environments—on when it is appropriate to waive signed written consent.

The Common Rule requirement for obtaining a signed written consent form may be waived under one of two conditions: (1) the signed form would be the only link of the participant to the research and the only risk would be disclosure of such participation; or (2) the research is minimal risk and is of a type for which written consent is not normally required outside of the research context (see Box 4-3). While minimal-risk biomedical research might not often qualify under the second exemption given that medical treatment usually requires people to sign a consent form except for the simplest procedures, minimal-risk SBES research could often qualify under this exemption. Many methods of SBES research involve activities that would rarely require signing a written consent form outside the research context (e.g., being observed in social activities by random bystanders, being asked to respond to a market survey, etc.). We review below types of survey, unstructured interview, observational, and secondary analysis research for which it is appropriate to waive signed written consent and note other consent issues such research poses.

Survey Research

Survey research using probability samples of households or individuals who represent a specified universe (e.g., the U.S. civilian noninstitutionalized population, people aged 55 and older, college graduates, or likely voters) has been a mainstay of SBES research in many disciplines since the 1940s. In order to obtain generalizable valid results, it is critical that a high percentage of sample cases respond to a survey. Sample cases who do not respond cannot be replaced without affecting the ability to make population estimates from the survey and to estimate the sampling error of those estimates. Furthermore, the failure to obtain responses from most sample cases may introduce systematic biases in the survey estimates for which reweighting and other methods do not compensate. These properties of survey methodology mean that recruitment procedures should be designed to inform prospective participants about the research, but the procedures should not raise unwarranted fears about the possible risks of participation or impose barriers to participation.

For this same reason, namely, to maximize response, investigators often provide financial incentives, such as cash or small gifts, to survey participants, and such incentives are generally appropriate. At a time when people have many demands on their attention and receive a large volume of unsolicited mail and telephone calls, it is becoming standard survey practice to recognize the burden on respondents by providing some type of reimbursement. Surveys may be lengthy, they may involve the cost of travel to the interview site or parking fees, or they may require special arrangements to accommodate family and work responsibilities such as babysitting costs. Normally, incentives include monetary reimbursement commensurate with respondents' opportunity costs for time and direct costs for travel, gift certificates, or small gifts such as vitamins or refreshments. The form and value of the incentive will vary, depending on the nature of the study and the potential burdens for respondents.

To facilitate survey research, we conclude that signed written consent should be waived, as a matter of standard practice, for minimal-risk mail, telephone, and in-person surveys of the general adult population that do not involve unusual incentives for participation and do not raise serious third-party consent issues (see "Third-Party Consent," above; see sections below for specific issues for risk assessment and consent in mail, telephone, and in-person surveys). Traditionally, survey research has not had a practice of obtaining signed consent for participation, and there is no evidence that the absence of written consent has jeopardized the rights or welfare of human participants. Indeed, the evidence reviewed above is that requiring written signed consent for minimal-risk surveys will discourage participation by some people who otherwise are willing to participate.

IRBs should carefully review survey protocols to be sure that the information that investigators provide to respondents by such means as interviewer scripts, advance letters, informational leaflets, and the like adequately fulfills the Common Rule requirements, particularly with regard to procedures for respecting privacy and protecting confidentiality. Documentation of consent to provide an audit trail can be assured in such ways as requiring that interviewers' presentation of scripts in telephone surveys are recorded or monitored; requiring that interviewers in personal surveys sign a statement that appropriate information was provided to respondents for their (tacit) consent; and calling back a sample of respondents to confirm that appropriate information was provided. Such procedures as callbacks and telephone monitoring are commonly used to validate that interviewers in fact administered the questionnaire and did not make up the information. It would be straightforward to add validation of the consent process to these procedures, if the protocol does not already provide for their use for this purpose.

Mail

Generally, mail surveys of the general population should be treated as minimal risk and, hence, not require written consent, even if the subject matter appears to be sensitive. The reason is that there is no interaction between the investigator and the participant and therefore no danger that a participant will feel threatened—the participant can simply toss the survey in the trash.[10] It is possible to imagine scenarios

---

[10] However, mail surveys of specific populations (e.g., cancer patients, people on therapy for HIV or AIDS) should have procedures to minimize the risk that respondents might be embarrassed or otherwise put at risk if someone other than the intended recipient opened the questionnaire package. Usually surveys to specially defined populations will be preceded by a letter or telephone call regarding participation and specifying procedures to protect privacy.

in which a respondent could have an adverse psychological reaction to a mail survey question. However, unless the study universe is known to include people who are likely to have serious adverse reactions to the question content (in which case a mail survey is probably not appropriate for the research in the first place), highly improbable scenarios should not drive the requirements for informed consent.

Because there is no opportunity in a mail survey to interpret information needed for informed consent (unless a respondent calls to ask for more information), it is critical that all materials mailed to respondents are appropriately targeted to the study universe in terms of reading level and language used. Because breach of confidentiality is almost always the only risk that could accrue to mail survey respondents, the survey materials should be clear about the level of confidentiality protection that will be provided and the purposes for which the respondent's data will be used (e.g., whether the data will be made available for any research purposes, including matching studies).

Telephone

Surveys conducted by telephone, unlike mail surveys, involve interaction between the interviewers and participants.[11] However, they are almost always minimal risk because the interaction is at arm's length, there is no intervention involved (the respondent is not subjected to any treatment), and the respondent may break off the interview at any time by hanging up the telephone. Thus, as for mail surveys, signed consent is rarely, if ever, necessary for participants in telephone surveys of the general population and for the same reasons—such documentation does not provide any added protection to the respondent, and it will likely reduce participation.

The interviewer's script should provide information the respondent needs in order to decide whether to participate, such as provisions for protecting confidentiality and the right of the respondent to refuse to answer questions and to break off the interview at any time. The IRB should determine that the interviewer script is understandable for the population of interest. The IRB should also establish that the investigator has procedures in place for quality control of interviewing, including a procedure to check that interviewers are adhering to the portion of the script that pertains to informed consent.

---

[11] Mail surveys often use telephone follow-up for people who did not mail back a completed questionnaire; the comments on telephone surveys apply to telephone follow-up as well.

In-Person

Surveys conducted in face-to-face interactions may represent minimal-risk or more-than-minimal-risk research depending on the nature of the study, the setting in which the research is conducted, and the vulnerability of the study population. Most often, written informed consent should be waived for in-person surveys involving minimal-risk research. In these situations, verbal consent is usually adequate. In studies involving more than minimal risk to the participants, written consent may be appropriate, but if research participants could come to harm because of potential stigmatization, emotional distress, or physical injury should there be a breach of confidentiality, it may be better to obtain consent verbally rather than create a paper record that could intentionally (e.g., by subpoena) or accidentally become public.

In many research studies involving face-to-face interviews, investigators must contact prospective participants prior to conducting the in-person survey. In these situations, investigators often send a letter requesting the individual's participation, indicating how he or she was identified, and describing the purpose, procedures, risks, and harm involved in the study, along with information about voluntariness of participation, protection of confidentiality, and incentives (if relevant). In some studies, it may be appropriate to have the initial contact made by a person known to the participant. For example, in a study involving face-to-face surveys with a population of student athletes, it would be appropriate for the athletes' coaches to send a letter or to verbally describe the study and inform the prospective participants whom they should contact if they are interested.

Unstructured or Semistructured Interviews

Unstructured or semistructured interviews are used by researchers conducting focus groups, oral histories, and some forms of ethnographic studies. They also may be used as one methodological tool in a study involving the application of several approaches to data collection. In some situations, interviews may be audiotaped and then transcribed. The primary characteristic of an unstructured interview is its allowance for informal discussion of particular topics.

Investigators who use unstructured or semistructured interviews confront unique ethical challenges when the study sample involves vulnerable populations or the interview addresses sensitive information about an individual. Deciding whether informed consent elements should be waived or whether to seek written or verbal consent requires a judgment based on the nature of the research, the population of interest, and the seriousness of the risks involved for participants in sign-ing an informed consent document. For example, verbal consent may be appropriate if participants are illiterate or vulnerable because of their legal status or involvement in illicit activities. Verbal consent may also be appropriate if the research is conducted in a cultural setting—nationally or internationally—in which signing a document to partici-pate in research is viewed as inappropriate.

Ethnographers often work in field settings in which they have fre-quent informal interactions with study participants. It would be bur-densome for both the researcher and the study participants to contin-ually obtain consent under these conditions. Verbal or written consent for semistructured interviews conducted during focus groups, oral his-tories, or ethnographic research should clearly identify strategies in place to protect the confidentiality of individuals. If audiotapes are used, individuals should be informed how the tapes will be stored, who has access to them, and when they will be destroyed or permanently archived. As in all cases, the process of informed consent should be evaluated against the standard of assuring participant respect and pro-tection and not adherence to a particular consent procedure.

Observational and Ethnographic Studies

Observational studies include those in which the investigator is ob-serving public situations when participants are anonymous and un-aware of the researcher, in which case consent is not relevant (see street-crossing observation example in Box 2-3 in Chapter 2), and those in which the researcher is known to the participants. Ethnographic re-search may involve direct, sustained observation of group behavior or the use of participant observation, in which the researcher is both a member and observer of a group. In both cases, the observation typi-cally involves a period of intense social interaction and engagement be-tween the ethnographer and individuals involved in the study. During this time, data (e.g., field note observations, interview results, archival materials) are systematically collected.

In ethnographic research involving direct observation of group ac-tivities, arrangements should be made before the implementation of the project to inform group members that the ethnographer will be present in the course of routine activities. In closed systems such as a hospital unit or an office or school setting, informed consent should be obtained from all those who are at the facility on a regular basis. Deciding whether to obtain written or verbal consent depends on the specific situation, including the vulnerability of the population being studied and the sensitivity of the information being collected. For example, in an ethnographic study of an intensive care unit in a hospital, verbal informed consent should be obtained from all staff members. Patients, their visitors, and other individuals who are not present on a regular basis but whose behavior may be observed in public activities should be alerted to the presence of an ethnographer if it is feasible to do so. Informed consent should always be obtained from individuals who are interviewed.

In some group observations, it may not be possible or necessary to obtain written or verbal informed consent from every person present. For example, at informal gatherings of visitors and staff at a nurse's station in a busy unit of a medical center, it would be intrusive to introduce the ethnographer, explain the study to every person who passed by, and obtain consent. In contrast, an ethnographer's presence at a family conference to discuss patient care and treatment decisions should be explained and permission obtained to observe the proceedings; if anyone is uncomfortable with the observation, it cannot proceed. If individual informed consent is going to be obtained from the participants in small group interactions, they should be advised about the methods designed to protect confidentiality.

Participant observation differs from other kinds of ethnographic research in that the researcher is a participant in the group—for example, a member of the local chapter of a political party in a study of grassroots politics—while at the same time the researcher observes the group. Often, other members of the group are not aware that they are being observed, and this technique raises a variety of ethical issues (see Wax, 1979; Bernard, 2000, 2001). For example, although the intent is to avoid altering the behavior of the group that might occur if the presence of an observer were known, it is possible that the behavior of the participant observer intentionally or unintentionally alters the behavior of other members of the group. If participant observation is designed to involve deception by not telling other members of the group that they are being observed, the IRB must decide whether it is permissible to waive some of the elements of informed consent (see "Omitting Elements of Informed Consent," below).

Analyses of Existing Data

Economists, sociologists, political scientists, psychologists, and other analysts often conduct research using existing data sets of individual records of variable size. When publicly available microdata sets are studied that were collected from research participants who gave consent for the original data collection for research purposes, no consent is required for secondary analysis. Indeed, IRBs should routinely exempt such research from review, given that the data have been processed using good practices to protect confidentiality (see Chapters 5 and 6).

Increasingly, secondary analysis involves abstraction and use of existing records as well as or instead of research microdata—for example, analysis of administrative records (e.g., state unemployment insurance records or food stamp case records), medical records, or academic records. Waiver of informed consent is appropriate when participants agreed to the use of their records for research when those records were originally collected and adequate confidentiality safeguards are in place. However, consent may not have been obtained at the time of collection, often because many records were collected in years past when research use of the data was not considered or when consent for such use was not a common practice. In some of these cases, it may be possible to obtain consent, and consent may be necessary to protect the study participants, so consent should be obtained. In cases when it is not possible to recontact participants to obtain consent, whether their records can be used without consent will depend on judgments about the researcher's ability to assure appropriate protection of the participants. If protection is believed to be adequate, then the records may be used. If protection is believed to be compromised, then the records may not be used. The decision must be made on a case-by-case basis. However, given evolving views about informed consent and increasingly complex issues surrounding confidentiality protection, greater effort must be made to find ways to respect and protect people whose records may be used for research in the future than was done in the past. Greater effort must be made to obtain consent at the time of original data collection.

OMITTING ELEMENTS OF INFORMED CONSENT

Recommendation 4.5: The Office for Human Research Protections should develop detailed guidance for IRBs and researchers, including specific examples, on when it is acceptable to omit elements of informed consent in social, behavioral, and economic sciences research.

The Common Rule acknowledges the appropriateness, under some circumstances, of omitting or modifying some of the required elements of the consent process. Although issuance of a waiver may seem to require extraordinary circumstances, the federal guidelines regarding waivers are quite straightforward. Some or all elements of informed consent may be altered or omitted if the research meets four criteria:

it is minimal risk; the waiver or alteration would not adversely affect participants' rights and welfare; the research could not practicably be carried out otherwise; and whenever appropriate, participants will be provided with additional pertinent information after participation (see Box 4-2, section d).

Furthermore, when a basic element of informed consent is clearly inapplicable to the proposed research, we argue that such an element should be omitted as a matter of common sense. For example, element (4) (see Box 4-1), which requires disclosing appropriate alternative procedures or courses of treatment, if any, that might be advantageous to the participant, is clearly oriented to biomedical or behavioral clinical research. Including that element in the consent form for, say, a study using structured or semistructured interviews would just add length and confusion to the form.[12]

Some SBES research may purposefully manipulate a condition of the research environment by deceiving participants about aspects of the research—passively by withholding information about the true purpose of the research, or actively by presenting information that is not correct in order to observe reactions to the condition created by that information. Deception should only be used when the welfare of participants has been carefully considered and judged to be protected and when other methods for studying the phenomenon of interest have been considered and judged not to be feasible.

Research involving deception is most often conducted in laboratory experiments in which the experimenter wants to create some belief or psychological state (deception is also sometimes used in participant observation and in field experiments; see the employment discrimination example in Box 2-2 in Chapter 2). In a conformity study, for example, a participant may be asked to write down a position and then be told that a specified proportion of other people in the room hold the opposite opinion. The intent of the research is to determine the effects of varying the proportion in opposition on the willingness of a participant to stick with his or her original position. The deception is judged not to put participants at risk of physical or psychological harm, and the study results could not be valid if participants were aware that the experimenter was manipulating the proportion of people with the opposing view.

---

[12]Our review of IRB websites at 47 major research institutions found that 28 percent of these IRBs do not indicate that statements about irrelevant elements of informed consent can be omitted from the consent document, and 9 percent require that statements about all of the basic elements be included in the consent document regardless of the nature of the research.

Because, historically, deception has been used in social psychological research it is not surprising that the American Psychological Association (APA) wrestled with issues of deception as early as the 1940s and 1950s. The topic took on increasing importance in the 1960s and 1970s in light of the controversy about such experiments as those conducted by Milgram and others (see Faden and Beauchamp, 1986:167-187; see also Box 3-1 in Chapter 3). In 1963, 38 percent of articles in journals in personality and social psychology reported uses of deception, as did 47 percent of articles published in the *Journal of Personality and Social Psychology* in 1971 (Faden and Beauchamp, 1986:172, 179).[13]

In 1973 the APA adopted a major revision of its code of ethics, which attempted to balance two concerns: the recognition "after almost 20 years of debate and self-study" that informed consent is a "moral ideal" for psychological research; and a view that the "strict application of informed consent would invalidate valuable research findings and would compromise the psychologist's ability to conduct meaningful research" (Faden and Beauchamp, 1986:185). The 1973 code was fairly general in its prescriptions. In the early 1990s, APA adopted a more explicit code in addition to a statement of principles. The code (revised in 2002, see http://www.apa.org/ethics [4/10/03]) essentially incorporates the Common Rule provisions: It explicitly and strictly limits the use of deception to research for which deception is necessary for valid results and the research is likely to have significant scientific, educational, or applied value; the research is not expected to cause physical pain or severe emotional distress; and participants are debriefed about the deception as soon as possible after data collection.

Discussion of deception needs to recognize that the term does not have a clear definition and can take on many meanings (see Smith, 1979), including deception about one or more aspects of a research protocol with the intention for full debriefing at the conclusion of the research; consent to participate in research in which participants are not fully informed; and consent to participate in research not knowing which of several treatments one will receive. Deception can also vary in the salience of the omitted or misleading feature, the extent of risk posed by the deception, and the extent of likely benefit from obtaining valid research results consequent on the deception.

It is not always clear where the boundary lies between deception and abbreviation of information about the study in order to reduce cognitive burden (e.g., informing prospective survey respondents of the main topics of the survey rather than showing them the entire ques-

---

[13]This journal specializes in research using deception designs; it is not typical of psychological journals.

tionnaire). Similarly, it is not always clear where the boundary lies between deception that is justified and necessary to produce valid research and deception that is so disrespectful of participants that it cannot be justified. In considering research protocols that involve one or another form of deception, IRBs and researchers should take care not to use the term pejoratively: deception is often well justified on scientific grounds, but, at the same time, a researcher should not make that decision in isolation.

The Common Rule and professional codes of ethics recognize the complexity of the issue and permit deception in appropriate circumstances, although the Common Rule restricts its use to minimal-risk research.[14] As with other aspects of informed consent (e.g., waiving written documentation or allowing proxy response for third parties), a key decision is the determination that a study is not of more than minimal risk, taking account of the research setting, method used, and population studied. For this reason, we recommend that OHRP provide guidance, with specific examples, to help IRBs make these judgments and to improve research practice.

CONCLUSION

Informed, voluntary consent is a critically important principle of human research participant protection and one of the most difficult to implement in an effective manner. We hope that a combination of systematic research on consent procedures and development of detailed guidance for IRBs and researchers will raise the standard of practice for seeking and documenting consent to participate in research in ways that increase the protection and respect for human participants and are commensurate with the risks of the research.

---

[14] It would be useful for IRBs, researchers, and participant representatives to debate the ethics of high-benefit research that requires deception for valid research results and is of more than minimal risk. Such research may be justified in some instances, but is unlikely to be approved given the current regulations and climate of fear in which IRBs operate.

— 5 —
Enhancing Confidentiality Protection

B REACH OF CONFIDENTIALITY, that is, the release of data that permit identifying an individual participant, is often the major source of potential harm to participants in social, behavioral, and economic sciences (SBES) research (see Sieber, 2001). For example, a survey that poses no risk of physical injury and no more than minor psychological annoyance to a respondent may nonetheless obtain data that could harm the respondent if others outside the research team (e.g., neighbors, co-workers, public agency officials) could associate those data with the person. Such information, if known by others, might affect employment, insurability, personal relationships, civil or criminal liabilities, or other activities or situations. In some cases, the simple fact of learning that an individual is a study participant could be harmful (e.g., if police or drug dealers were to learn the names of participants in an ethnographic study of drug markets). Furthermore, if a participant has been assured of confidentiality, then disclosure of identifiable information about the person is a violation of the principle of respect for persons even if the information is not sensitive and would not result in any social, economic, legal, or other harm.

Protection of confidentiality is a concern in SBES research whenever data are collected in identifiable form. Identifiers include not only such overt information as name, address, social security number, telephone number, and e-mail address, but also detailed information about the respondent, such as income and profession, that could permit identification by inference in the absence of an explicit identifier.[1] Some SBES research does not collect identifiable information in these terms—for example, observational studies of street-crossing behavior of people who are not photographed or approached by the investigator in any way. However, for much SBES research, confidentiality protection is a necessary and vitally important component of the study plan.

---

[1] Even the assignment of arbitrary identifiers may not protect against re-identification so long as the link between the arbitrary codes and originally collected real identifiers (e.g., name) has not been destroyed.

Breach of confidentiality can occur at any stage of a research project—data collection (including recruitment of participants), processing, storage, and dissemination for secondary use. At the present time, the risk of disclosures that could be embarrassing or damaging to participants (or that could simply violate a pledge of confidentiality) is increasing because of several factors. Most of these factors affect the disclosure risk for dissemination for secondary use, but some also have implications for the disclosure risk as a result of data collection, processing, and storage. They include the following:

- There are growing numbers and variety of publicly available microdata files for secondary analysis. Such files provide information on individuals that have been stripped of obvious (and less obvious) identifiers. Increasingly, microdata sets contain richly detailed content from multiple observations on the same individuals over time (panel surveys), or they contain data on more than one type of entity (e.g., education surveys of students, their parents, teachers, and schools), or they contain both kinds of data. Such rich data sets increase the potential for re-identification of respondents through linkages with other data sources. Panel surveys also pose disclosure risks as a result of data storage because contact information must be retained for respondents for months or years. Generally, disclosure risks for panel surveys increase over time.

- There are growing numbers, variety, and content of administrative records data sets from public and private agencies (e.g., birth and death records) that are readily available on the Internet. Such files can potentially be linked to research data sets and used to re-identify research respondents.

- More broadly, the capabilities to link information across multiple sources on the Internet are increasing.

- There is increased emphasis by funding agencies on data sharing among researchers, not only to permit replication of results, but also to foster additional research at low marginal cost. Such sharing has many benefits, but it also multiplies the number of people with access to the data.

- The speed of data processing and volume of low-cost data storage are increasing, which facilitates efforts to link data sets.

- There is increased use of data collection technologies, such as web surveys, and data transmission methods, such as e-mail and file-sharing procedures, that may not be secure.

In this chapter we provide historical background on confidentiality protection for research data in the United States, beginning with the attention given to protection issues by the U.S. Department of Health and Human Services (DHHS) and the institutional review board (IRB) system. We continue with the history of legislative protection for data collected by the Census Bureau and other federal statistical agencies that are widely used by SBES researchers and others. (For decennial census data, legislative protection goes back to the 1920s.) Until fairly recently, the activities of IRBs and statistical agencies with regard to confidentiality protection have proceeded largely independently.

Next, we provide a fuller explication of the factors that are challenging the adequacy of confidentiality protection measures today and the techniques and procedures that statistical agencies are adopting in response. Our recommendations to IRBs, the Office for Human Research Protections (OHRP), and research funding agencies for enhancing confidentiality protection for different kinds of SBES research follow. To protect participants and facilitate research with existing data, we propose a new system for certifying the confidentiality of data files, built on existing and new data archives in the United States.

HISTORY OF CONFIDENTIALITY PROTECTION IN THE PARTICIPANT PROTECTION SYSTEM

Common Rule and IRB Operations

Surprisingly, the history of human research participant protection policies and regulations shows relatively little attention to issues of confidentiality protection.[2] Although the 1966 U.S. Public Health Service policy statement and the 1971 "Yellow Book" guidelines[3] for participant protection mentioned the need to protect confidentiality, the 1974 regulations (45 *CFR* 46) did not require IRBs to determine that study plans adequately address confidentiality issues. Indeed, papers and testimony from social scientists prepared for the 1974-1978 National Commission for the Protection of Human Subjects of Biomedical and Behavioral Research commented that existing regulations did not adequately address issues of confidentiality in SBES research. A provision

---

[2] By "confidentiality," we mean protecting private information from being revealed to others in a way that could identify an individual research participant. Such protection is distinct from "privacy," by which we mean the right of an individual to decide whether to share information with the investigator in the first place (e.g., a survey participant could refuse to answer certain questions on grounds that they invaded his or her privacy; see National Research Council, 1993:22-23).

[3] The "Yellow Book" was the name given to "The Institutional Guide to DHEW Policy on Protection of Human Subjects" (see Chapter 3).

on confidentiality was added to the 1981 version of the regulations (45 *CFR* 46.111a): it required IRBs to determine "where appropriate, there are adequate provisions to protect the privacy of subjects and to maintain the confidentiality of data." The 1981 regulations also specified "a statement describing the extent, if any, to which confidentiality of records identifying the subject will be maintained" as one of the basic elements of informed consent (45 *CFR* 46.116a). Beyond these two references, however, the Common Rule provides no guidance, even on traditional confidentiality protections for laboratory, survey, ethnographic, and other originally collected data, such as assigning new identifiers and destroying the link to the original identifiers, keeping data records in locked files, and the like.[4] Guidance in the *IRB Guidebook* (Office for Protection from Research Risks, 1993:Ch.III.D) is very general.

With regard to IRB attention to confidentiality protection in reviewing protocols, the 1975 Michigan survey found that confidentiality issues were relatively rare as a focus of IRB review: only 3 percent of protocols were required to modify their confidentiality procedures; in comparison, 24 percent of protocols were required to modify their consent forms or procedures (Gray, Cooke, and Tannenbaum, 1978:Table 2). This difference may be understandable given that the challenges to maintaining confidentiality were not as great then as they are today. However, the 1995 Bell survey 20 years later reported similar results: only 3 percent of IRB chairs said that inadequate confidentiality protections were often a problem with the research protocols they reviewed, while problems with consent forms were cited frequently (Bell, Whiton, and Connelly, 1998:Figure 40). Similarly, only 14 percent of investigators reported being required to modify their procedures for protection of privacy and confidentiality, compared with 78 percent who reported being required to modify their consent forms (Bell, Whiton, and Connelly, 1998:Figure 41).

The continued relative lack of emphasis on confidentiality protection may result from determinations by IRBs that proposed protection procedures are adequate. It may also result from continued underestimation by IRBs of the risks of disclosure, which today's research and computing environment has heightened.

---

[4]See Sieber (2001) for a critique of the Common Rule's limited statements on confidentiality, which she asserts do not properly recognize the distinction between confidentiality and privacy.

Confidentiality Certificates

Another initiative by federal research funding agencies to protect confidentiality is the long-standing program of the National Institutes of Health (NIH) whereby researchers may obtain certificates of confidentiality for research on sensitive topics, whether the research is funded by NIH or another agency. The National Institute of Justice also makes such certificates available for criminal justice research. These certificates buttress confidentiality protection in specific circumstances—namely, they protect researchers from being compelled to deliver names or identifying characteristics of participants in response to court orders or subpoenas, unless respondents have consented to such release.[5] Qualifying studies include those that collect data on such topics as sexual attitudes, preferences or practices; use of alcohol, drugs, or other addictive products; mental health; genetic makeup; illegal conduct; or other topics for which the release of identifiable information might damage an individual's financial standing, employability, or reputation within the community or might lead to social stigmatization or discrimination. At present, however, the protection afforded by such certificates is prospective; that is, researchers cannot obtain protection for study results after data collection has been completed, and it is not always obvious in advance when a certificate may be needed.

Medical Records Protection

The Health Insurance Portability and Accountability Act (HIPAA) of 1996 contained a provision that has resulted in the latest initiative by federal research funding agencies to protect confidentiality. HIPAA promotes the use of standard formats for electronic information exchange to simplify the administration of health insurance payments for medical treatment. Recognizing a potential threat to the confidentiality of patient records, HIPAA required DHHS to submit to Congress detailed recommendations on privacy standards for individually identifiable health information. This short provision led to the Privacy Rule, which comprises hundreds of pages of regulations and commentary; it is scheduled to take full effect in April 2003 (see Gunn et al., 2002; Institute of Medicine, 2002:205-211).

The version of the Privacy Rule issued by DHHS in December 2000 drew substantial criticism from the health care community, including researchers, who complained that the provisions for research access

---

[5]The New York Court of Appeals upheld the authority of confidentiality certificates in 1973 (for more information, see http://grants1.nih.gov/grants/policy/coc [4/10/03]).

to health information were confusing and unnecessarily restrictive. In response, DHHS published a modified Privacy Rule in August 2002.

The Privacy Rule applies to health plans, health care clearinghouses, and health care providers (covered entities) who maintain patient and claims records; it also affects health care researchers who obtain such records from covered entities for analysis purposes. Under the rule, covered entities may make "de-identified" data available for research use, without patient authorization, in one of three ways. First, a covered entity may release a "limited data set," consisting of patient and claims records stripped of a list of direct identifiers of the individual, relatives, household members, and employers, to researchers who sign a legally binding agreement to safeguard and not disclose the information. The identifiers that must be deleted include names, street addresses, telephone numbers, e-mail addresses, social security numbers, medical record and health plan account numbers, device identifiers, license numbers, vehicle identifiers, full face photos, and finger and voice prints (Gunn et al., 2002:8). However, birth date, 9-digit zip code, and dates of admission and discharge are permissible to include in such a data set.

Second, a covered entity may release a "de-identified" data set for research use without requiring the researcher to sign an agreement provided a more comprehensive list of identifiers has been removed. Third, a covered entity may employ a statistician to attest that the risk of re-identification is very small because of the nature of the data (e.g., in cases when the data have been subject to statistical manipulation—see "Protection Methods of Statistical Agencies," below).

The Privacy Rule also provides that IRBs or Privacy Boards may issue waivers for research access to data when the research cannot be conducted with de-identified data and when it is not practicable to obtain authorization from research participants.[6] The waiver requirements were initially criticized as being inconsistent with the Common Rule; they were simplified and rewritten for consistency. They require that adequate plans are in place to protect identifiers from disclosure and to destroy them at the earliest opportunity. IRBs and Privacy Boards may also allow researchers limited access to identifiable records data in order to identify and recruit prospective participants or to conduct preliminary exploratory research to determine the feasibility of a full-fledged analysis.

We cannot do justice to the Privacy Rule provisions in this brief summary nor anticipate how they may work in practice. We note

---

[6]It is apparently expected that IRBs would handle waivers under the Privacy Rule for federally funded research and that Privacy Boards would handle waivers for other research, although this is not clear (see Institute of Medicine, 2002:209).

that these provisions increase the necessity for IRBs, OHRP, and researchers to become cognizant of good practices for confidentiality protection, as discussed below.

CONFIDENTIALITY PROTECTION IN THE FEDERAL STATISTICAL SYSTEM

Census Bureau History

1790 to World War II

The history of confidentiality protection for federal statistical data begins, as for so many data collection and dissemination issues, with the decennial census—first conducted in 1790 pursuant to the U.S. Constitution.[7] In the first few decades of the census, the returns were posted in public places for public review. By the middle of the 19th century, the Congress and census directors began to worry about enumerators improperly revealing information and possibly gaining some private benefit. Public posting was discontinued, and enumerators were instructed to keep census information confidential. Yet federal, state, and local agencies and courts not infrequently attempted to obtain individual census returns. Most often the Census Bureau rebuffed these requests, but sometimes it acceded to them. Finally, Public Law 13 (Title 13 of the U.S. Code) was enacted in 1929 to codify various practices that had been emerging in official U.S. statistics. Section 9 explicitly provided for the confidentiality of economic and population census data:

> The information furnished under the provisions of this Act shall be used solely for the statistical purposes for which it is supplied. No publication shall be made by the Census Office whereby the data furnished by any particular establishment or individual can be identified, nor shall the Director of the Census permit anyone other than the sworn employees of the Census Office to examine the individual reports.

Another section provided heavy penalties, which currently include large fines and up to 5 years' imprisonment, for Census Bureau employees who breach confidentiality. Title 13 also covers household surveys conducted by the Bureau that use the census address list as their sampling frame.

---

[7]This history of confidentiality protection for the U.S. census draws heavily on Gates (2000) and Seltzer and Anderson (2002).

The enactment of Public Law 13 was timely because the Census Bureau was publishing more and more tabulations for smaller and smaller geographic areas, which required careful specification and review to minimize the risk of individual identification. As early as 1910, the Bureau published data for census tracts (locally delineated neighborhoods) in selected cities that paid for the tabulations. By 1940 the Census Bureau was coding and publishing census tract data for 64 cities. Also in 1940 the Bureau introduced a program of statistics for individual blocks in 191 cities.

World War II and Later

At the outbreak of World War II in 1939, the U.S. Attorney General sought legislation to amend Title 13 to allow military and intelligence agencies to have access to individual census records. The Census Bureau adamantly opposed the legislation, and it was withdrawn. However, in June 1941 a newly appointed Census Bureau director, J.C. Capt, obtained the support of the Commerce Department for legislation to authorize periodic surveys for national defense needs and to make census reports for individuals available for use in the "national defense program" with the approval of the president. This legislation passed the Senate in August 1941 with an accompanying report (77th Congress 1st session, Senate Rept 495, June 26, 1941, to accompany S 1627) that said:

> The needs of the defense program are of such a character as to require full and direct information about specific individuals and business establishments. . . . To continue to impose the rigid provisions of the present confidential use law of the Census Bureau. . . would defeat the primary objects of the legislation here proposed.

The Senate legislation did not pass the House, but the Second War Powers Act, enacted March 27, 1942, effectively incorporated its provisions. This act provided that any Department of Commerce data could be provided to any federal agency at the written request of the agency head. It is not known whether individual census reports were ever provided to people other than sworn Census Bureau employees. However, census tract-level tabulations of Japanese Americans from the 1940 census were provided to the Office of Naval Intelligence, and maps of city blocks with counts of Japanese Americans were provided to the Western Defense Command of the War Department, which facilitated internment of legal residents of Japanese origin.

The relevant section of the Second War Powers Act was repealed as part of the First Decontrol Act of 1947. In 1947 the Census Bureau refused a request by the Attorney General for census information on individuals who were suspected of being communist sympathizers. Since that time, the Bureau has an unblemished record of protecting confidentiality for the data it collects from respondents to censuses and surveys, despite the increasing challenges it faces to such protection.[8] Its standing Disclosure Review Board reviews every data product the Bureau makes available for public use to ensure that disclosure risks are minimized.

Other Statistical Agencies

All federal statistical agencies operate under strong norms to protect the data they collect against disclosure that could identify an individual.[9] Some agencies have legal protection against requests from administrative agencies and other bodies to disclose individually identified information. However, other agencies have had to rely on executive orders, court cases, and long-established custom (see Norwood, 1995).

For years, the Statistical Policy Division of the U.S. Office of Management and Budget (OMB) endeavored to obtain legislation that would strengthen the statutory basis for protecting the confidentiality of all federal data collected for statistical purposes under a confidentiality pledge. These efforts achieved success when, in November 2002, Congress enacted the E-Government Act of 2002. Title V, the "Confidential Information Protection and Statistical Efficiency Act of 2002," subtitle A, places strict limits on the disclosure of individually identified information collected under a pledge of confidentiality: such disclosure can occur only with the informed consent of the respondent and the authorization of the agency head and only when the disclosure is not prohibited by any other law (e.g., Title 13). Subtitle A also provides penalties for employees who unlawfully disclose information (up to 5 years in prison, up to $250,000 in fines, or both).

However, even though confidentiality protection for statistical data is now on a much firmer legal footing across the federal government, a loophole may exist for data from the National Center for Education

---

[8] See Gates (2000) for a summary of post-World War II changes in legislation and court decisions that have upheld the confidentiality protections of Title 13. A legal exception to Title 13 is the provision in Title 44 that allows the National Archives to obtain individually identified census records and make them available for research use 72 years after the census date.

[9] See *Principles and Practices for a Federal Statistical Agency* (National Research Council, 2000b).

Statistics (NCES). NCES has for many years had strong statutory protection for maintaining the confidentiality of its data and stiff penalties for NCES staff who breach confidentiality. The USA Patriot Act of 2001, enacted in October 2001 following the tragic terrorist acts of September 11, may have vitiated the legal protections for NCES data. Section 508 of the act amended the National Center for Education Statistics Act of 1994 by allowing the Attorney General (or an assistant attorney general) to apply to a court to obtain any "reports, records, and information (including individually identifiable information) in the possession" of NCES that are considered relevant to an authorized investigation or prosecution of domestic or international terrorism. Section 508 also removed the penalties for NCES employees who furnish individual records under this section. To date, no requests for such records have been made, but NCES is revising the information it provides to survey respondents about the possibility that their data could be obtained under this act. It is not yet clear whether the confidentiality protections in the E-Government Act would take precedence over Section 508 of the Patriot Act.

Federal Statistical Agencies and IRBs

Most but not all cabinet departments that house federal statistical agencies have formally adopted the Common Rule (exceptions are the U.S. Departments of Labor and Treasury), and agency IRBs review proposed surveys for many statistical agencies. For example, the National Center for Health Statistics has an IRB, and the IRB for the Department of Education reviews NCES surveys. The Census Bureau, in contrast, does not obtain IRB review on the basis that its surveys are exempt under 45 *CFR* 46.101(b)(3)(ii). That provision exempts research from IRB review when federal law, as in Title 13, requires "without exception that the confidentiality of the personally identifiable information will be maintained throughout the research and thereafter." Yet there are features of some Census Bureau surveys that might be viewed as requiring IRB review (e.g., the appropriateness of providing financial incentives only to cases that otherwise refuse to participate in the Survey of Program Dynamics).

All federal surveys are subject to clearance by OMB under the provisions of the Paperwork Reduction Act. This review covers not only survey costs and burden for respondents, but also such issues as whether respondents are adequately informed about the purpose of the survey, the use of the information, whether response is voluntary or mandatory, and the nature and extent of confidentiality protection. We are not in a position to recommend whether IRB review is needed in addition to OMB review, but we do suggest it might be useful for OMB and OHRP to discuss their respective jurisdictions. We note that statistical agencies have encountered some of the same problems as SBES researchers with IRB review, such as insistence on requiring signed written consent for minimal-risk surveys when evidence indicates that a signature requirement will deter response from some people who would otherwise be willing to participate (see Chapter 4).

PROTECTING CONFIDENTIALITY TODAY

Increasing Challenges

The development of new data collection and dissemination technologies is arguably the principal factor increasing disclosure risks for research data that are made available by federal statistical agencies and other providers today. Other factors that play a role include increases in the volume and richness of the data collected (in turn made possible by technological advances) and changes in the nature of SBES research, which increasingly involves secondary analysis of data collected by others and sharing of data for validation purposes.

New Technology

Collection and processing technology for large-scale data collection efforts has been under almost continuous development since at least the end of the 19th century, when Herman Hollerith (then a Census Bureau employee, later, the founder of IBM) invented a punch-card tabulation machine to edit and tabulate the 1890 census (see Salvo, 2000). At that time and for many years thereafter, the limitations of printing technology constrained the amount of tabulations that the Census Bureau and other agencies could publish for research use, thereby minimizing disclosure risk.

The challenges of protecting data confidentiality began increasing in the 1960s when the Census Bureau first took advantage of computerization to greatly expand the volume and kinds of data it made available to the user community. From the 1960 census (the first to be processed wholly by computer), the Bureau provided summary files (SFs) for small geographic areas on a reimbursable basis to several business firms. The tabulations on these computer files were much more extensive than those in printed reports. In 1963 the Bureau, with support from the Population Research Council, developed the first public-use microdata sample or PUMS file, which contained 1960 census individual records for 180,000 people (a 1-in-1,000 sample of the U.S. population).[10] In the 1970 and subsequent censuses the Bureau greatly expanded the SF and PUMS programs, and, beginning in the late 1970s, the Bureau and other statistical agencies provided public-use microdata files for an increasing number of large household surveys, including the Consumer Expenditure Survey, Current Population Survey, and Health Interview Survey.

Yet throughout the 1970s threats to confidentiality were lessened by the small number of secondary users and the difficulties of acquiring and working with large computer files from the Census Bureau and other statistical agencies. The files were generally expensive to acquire (even though users only had to pay the costs of reproduction); they were also expensive to process, requiring programming support, mainframe computer hardware, and, often, investment in customized software. Users also required considerable training in how to analyze and interpret the data. Hence, the barriers to use were high.

The spread of personal computing in the 1980s and 1990s greatly expanded the number of users who conducted secondary analyses of summary and microdata files from statistical agencies and other sources. However, at least initially, the storage capacity and processing speeds of personal computers were limited, thereby limiting the ability for linkage across data sets or other types of data manipulation that might breach confidentiality.

In the 1990s, the emergence of the world wide web, together with vastly increased computing power and storage capacity of personal computers (often networked to provide yet more capacity), began to markedly increase the potential for breaches of confidentiality to occur. Multiple data sets were made available on the web, including summary and microdata files from statistical agencies and records of various types from public and private agencies. The volume of easily accessible data, together with sophisticated matching software, increased the likelihood that a determined investigator might re-identify a survey respondent despite the best efforts of individual agencies to minimize disclosure risk.

Paralleling developments in data processing and dissemination technology were developments in technology for data collection from survey respondents. Computer-assisted telephone interviewing (CATI) came into use beginning in the 1980s, followed by computer-assisted personal interviewing (CAPI) in the 1990s. CAPI technology, in particular, in which interviewers record responses on laptop computers

---

[10]We define a public-use microdata file as a computer-readable file that contains individual records for a sample of individuals or households, is intended for research use, is available to any user, and has been processed to minimize the risk of identifying a particular individual by using widely recognized good practices for confidentiality protection.

in the field and transmit the data over telephone lines to agency head-quarters, posed new problems of protecting confidentiality at the stage of interviewing and data transmission. Most recently, survey organizations and individual researchers have experimented with collecting responses on the Internet, which poses yet more challenges for confidentiality protection.

Data Richness

The kinds of technological developments discussed above, including faster processors, more storage capacity, and more sophisticated data processing and analysis software, made possible spectacular growth over the past three decades in the volume and richness of data sets that are available for secondary analysis by SBES researchers. This growth was also fueled by increasing demands for applied research in such areas as health services, retirement behavior, education, work and welfare, which have been the focus of public attention and policy debate.

In terms of sheer volume of observations, PUMS files containing microdata from the decennial census long-form sample expanded over this period from a 1-in-1,000 sample of the population in 1960, totaling more than 180,000 records, to as large as a 1-in-20 sample of the population in 2000, totaling more than 10 million records. Microdata files from the major household surveys of statistical agencies are also large and complex: for example, the Current Population Survey March Income Supplement contains data for more than 70,000 households and 180,000 people with detailed information on employment, family income, and household composition.

Even more exciting from the perspective of SBES research has been the development of complex longitudinal, multilevel surveys that provide a vast array of information to support in-depth secondary analysis of specific populations. Some of the major surveys of this type are the Health and Retirement Survey, National Longitudinal Survey of Mature Women, National Longitudinal Survey of Youth, Panel Study of Income Dynamics, Survey of Income and Program Participation, and Survey of Program Dynamics. To illustrate the breadth and depth of information such surveys can contain, Box 5-1 summarizes the structure and content of the Health and Retirement Survey, which is conducted by the University of Michigan Survey Research Center with a grant from the National Institute on Aging.

BOX 5-1
Health and Retirement Survey Design and Content

Design

The first cohort began in 1992 with 12,654 men and women aged 51-61; the second cohort began in 1998 with a smaller sample size. For both cohorts, they are interviewed every 2 years (spouses are also interviewed). New cohorts are to be introduced every 5 years.

Linkages are performed or planned with Medicare records; social security earnings records; National Death Index; employer health plans; and employer pension plans (summary plan descriptions).

Content (First Cohort)

Not all questions were necessarily asked in every interview. The questionnaire also includes job history, income, and demographic characteristics, in addition to the topics listed below.

Retirement-Related Expectations (for employed people)
  Probability of being laid off in next year
  Probability of finding an equally good job if laid off
  Whether would accept move to another state or a layoff
  Probability of working full-time after age 62, after age 65
  Probability that health will limit activity during next 10 years
  Expect real earnings to go up, down, or stay the same in next few years
  Retirement plans: whether expect to retire completely, never stop work, work fewer
       hours, change kind of work, work for oneself, haven't thought about it
  How much personal savings expect to have accumulated by time retire
  Whether and how much expect living standards to change after retirement

Other Probabilities, Expectations
  Whether expect to have to give major financial help to family members in next 10
       years
  Whether will live to age 75 or more, age 85 or more
  Whether housing prices in neighborhood will go up faster than prices in general
       over next 10 years
  Whether Congress will make social security more or less generous
  Whether U.S. will experience major depression in next 10 years
  Whether U.S. will experience double-digit inflation in next 10 years
  When expect to receive social security, how much in today's dollars, ever had SSA
       calculate benefits
  Looking 2 years ahead, whether expect to be better or worse off financially

Risk Aversion, Time Preference
  Whether would take another job with 50-50 chance it would double family income
       or cut by a third; with 50-50 chance it would double family income or cut in
       half; with 50-50 chance it would double family income or cut by 20 percent.
  In deciding how much to spend or save, which time period is most important: next
       few months, next year, next few years, next 5-10 years, longer than 10 years

Attitudes Toward Bequests
  Importance of leaving a bequest
  Whether expect to leave a sizable bequest

BOX 5-1 (continued)

Self-Reported Pension Coverage on Current Job (Similar questions are posed for the previous job if respondent is not working and for each job in job history section)

    If participating, for each plan, whether defined benefit or defined contribution or combination

    For each defined contribution, type of plan, how much accumulated, how much employer contributes, how much respondent contributes, how many years in plan in total, whether can choose how money is invested and whether mostly stock or interest-earning assets or evenly split, whether can receive lump sum or installments, youngest age when could start receiving benefits, what age expect to receive benefits and in what form

    For each defined benefit, age for full benefits and how much, expected earnings at full retirement age with this employer, age for reduced benefits and how much benefits would be reduced, whether plan benefits depend on social security benefits, whether can take lump sum

    If not participating, whether employer offers pension plans, whether respondent eligible and intends to participate in future and whether employer contributes

Heath Status

    Self-reported health status now and compared with a year ago

    Self-reported emotional health status

    Difficulty with activities of daily living, including instrumental activities

    Self-reported medical conditions indicated by a doctor (high blood pressure, diabetes, cancer, chronic lung disease, strokes, emotional problems, arthritis, other problems, broken bones, pain, poor eyesight, hearing problems)

    Self-reports of smoking, drinking, exercise

    Cognition battery and mood assessment and clinical depression battery

    Self-reported work disabilities and employer accommodations

Health Insurance Coverage

    Type of coverage: government, employer, individual, other

    If employer coverage, whether employee pays part or all of premium, whether available to retirees and whether employer pays part or all, whether retirees pay the same as other employees, whether spouses can be covered and whether retirees pay the same for spouse coverage as other employees

    If individual coverage, type and cost

    Whether ever turned down for coverage and why

Health Care Use and Costs

    Stays in hospital or nursing home last 12 months

    Doctor visits last 12 months

    Home health care last 12 months

    Itemized medical care deductions

    Cost of individual insurance

    Total and out-of-pocket medical care expenditures, by category of service

Assets and Debts

    Value of house, mobile home and site, farm, ranch

    Amount of mortgage, second mortgage, home equity loan

    Value of second home, time-share, amount of mortgage

    Net value of motor home or recreational vehicle

    Net value of other real estate, other vehicles, business

    Amount in Individual Retirement Accounts or Keogh accounts

    Net value of stocks, mutual funds

    Money in checking, saving, and money market accounts

BOX 5-1 (continued)

Money in certificates of deposit, government savings bonds, Treasury bills
Money in corporate bonds
Net value of other savings/assets
Amount of other debts
Inheritances, when and from whom received, worth at the time
Value of other transfers of $10,000 or more from relatives
Life insurance settlements, when received, worth at the time, who was insured
Large, unexpected expenses over last 20 years that made it difficult to meet financial goals
Cash value of life insurance
Capital gains component of asset value increases after first interview
Expenditures (see above for medical care)
Mortgage, rent, taxes, utilities, condominium fees
Financial assistance of $500 or more in past 12 months to children or parents
Food per week or month (including value of food stamps) in stores and delivered
Meals eaten out (not counting at work or school)
Itemized medical care deductions
Charitable contributions (if $500 or more)
Support to others outside household
Total expenditures in third wave

SOURCE: National Research Council (1997:Table 4-1, Boxes 4-4 to 4-7).

At the same time, there has been an expansion in the volume and richness of administrative and other data sets collected by public and private agencies (see Sweeney, 2001). For example, birth certificates in many states now contain much more information than previously. The private sector has developed vast files of customer preferences and shopping habits. Although federal and state agencies have developed rules for data access and confidentiality protection for many public sector administrative data sets, some public and private sector data on individuals are accessible on the web. This development increases the opportunities for linkage with research data sets and increases the need to develop innovative confidentiality protection measures that minimize disclosure risk while not so restricting or altering the data as to undercut their research usefulness.

SBES Research Environment

Changes in the SBES research environment have increased the risks of disclosure and the need to pay heed to confidentiality protection. As a result of technological developments in data processing, dissemination, and analysis, and the increased richness, variety, and volume of microdata sets, large numbers of SBES researchers engage in secondary analysis. These researchers include labor and welfare economists, health services researchers, sociologists and social psychologists, educational researchers, cultural anthropologists, and public opinion researchers. They obtain PUMS and summary files not only directly from source agencies, but also, increasingly, from data archives housed at universities that acquire files for redistribution from federal agencies, researchers, and others. Such archives include the Interuniversity Consortium for Political and Social Research (ICPSR) at the University of Michigan, which provided researchers at member universities access to public-use microdata as far back as the early 1960s (in the form of punchcards); the University of California (Berkeley) Data Archive; the University of Minnesota Population Resource Center; the University of Wisconsin (Madison) Data and Program Library; and many others.

The growth in secondary analysis has whetted appetites for ever richer data sets, including linkages of survey microdata with such administrative data as social security earnings records, vital statistics records, Medicare and Medicaid records, employment and public assistance records from state and local agencies, and employer benefit records (see Hotz et al., 1998). These kinds of linkages can be difficult to achieve, given that custodial agencies (federal and state agencies, employers) generally have their own rules for access, which can differ for the same type of data among agencies. Yet once achieved, such linkages raise disclosure risks if the researcher does not take adequate measures to protect confidentiality.

The interests of researchers in access to rich data sets has often resulted in an adversarial stance with data providers, particularly statistical agencies. Researchers are often impatient with agency restrictions on data access, yet they often do not fully understand the restrictions under which such agencies operate. Not only would statistical agencies be subject to stiff penalties if researchers were able to re-identify respondents in their data products, but they would also be concerned that confidentiality breaches would destroy trust with respondents and make it harder to obtain high enough response rates for quality results.[11]

Finally, the policies of research funding agencies and leading academic journals are pushing researchers to share their data with others. For example, the NIH Office of Extramural Research recently updated its policy guidance to expect researchers to share data from NIH-supported studies on a timely basis for use by other researchers. Most investigators submitting an NIH application will be required to

---

[11] Surveys have shown that many people do not trust the confidentiality pledges from statistical agencies and that such mistrust can lead to reduced response (see, e.g., Singer, 2001).

include a plan for data sharing or to state why data sharing is not possible (see http://grants.nih.gov/grants/policy/data_sharing [4/10/03]). For many years, the National Science Foundation (NSF) economics program has required data underlying an article arising from an NSF grant to be placed in a public archive. Similar expectations exist at the National Institute of Justice and the Robert Wood Johnson Foundation. Moreover, many scientific journals require that authors make available the data included in their publications.

The interest of funding agencies is primarily in leveraging their investment in data collection by encouraging investigators to share data with others for secondary analysis. The interest of journals is primarily in being able to assure the quality and validity of the researcher's findings by making it possible for others to replicate the results. (See National Research Council, 1986, for a discussion of the benefits of sharing research data.)

Researchers who are conducting secondary analysis of publicly available summary or microdata would have no problem in satisfying requirements for data sharing. However, the many researchers who collect their own data or who conduct analyses with data obtained from a variety of sources (e.g., linked survey and administrative data) could find it difficult to determine how to share data with others in a way that does not increase disclosure risks.

Protection Methods of Statistical Agencies

Because their mission is both to provide data for public use and to ensure that individual respondents are not re-identified, federal statistical agencies are leaders in the development of techniques and policies for confidentiality protection at every stage of data development (see Doyle et al., 2001). Sometimes they have reacted to the heightened risks of disclosure from the technological and other developments just discussed by curtailing the availability of data. We briefly review some of the confidentiality protection practices for data dissemination of the Census Bureau and other federal statistical agencies to illustrate the range of techniques and policies used. The paper by George Duncan in Appendix E discusses disclosure risk analysis in detail and describes a range of methods for processing summary and microdata files to protect confidentiality.

Census Bureau

The Census Bureau, because it collects and distributes such large volumes of data and because of the strict confidentiality protection provisions of Title 13, has been proactive in developing techniques to minimize the risks of re-identifying respondents from its data products, as well as to ensure confidentiality at the stages of data collection and processing. For example, an option for responding to the 2000 census for households that received the short form in the mail was to answer the questions over the Internet. Respondents had to enter the 23-digit control number on the mail questionnaire to authenticate their response and preclude duplicate responses. Their data were entered through a firewall on the Bureau's website, encrypted, sent to the Bureau's main computer center, and put behind a second firewall to protect confidentiality.

Every data product the Bureau makes available for public use must be reviewed by its Disclosure Review Board to ensure that disclosure risks have been minimized. For summary (tabular data), the Bureau's measures to protect confidentiality include several steps. The level of detail of tabulations for geographic areas is related to the size of the area and the size of the survey or census sample.[12] For census tabulations, the Bureau uses a "data swapping" technique (a type of data blurring), in which a small number of records for individual households that are similar on basic characteristics (number of adults and children and race and ethnic composition) are swapped between adjacent geographic areas so that the resulting tabulations for individual areas are close to but likely not exactly the same as the originally collected data.[13] The Bureau also groups reported amounts for such continuous variables as income, rent, and housing value into broad categories, including a top category that is well below the largest individual amount reported.

For microdata files of individual records that are made publicly available on the Internet or other media (PUMS files), the Bureau protects confidentiality by taking such steps as stripping off all overt tags, such as name and address; limiting geographic identification to large areas (e.g., states, regions, or metropolitan areas above a certain population size, depending on sample size); top-coding continuous variables (e.g., providing income amounts in dollar increments up to a top category defined as any income that exceeds a specified amount); and assigning such variables as specific occupation, industry, or an-

---

[12] For example, only basic census characteristics collected from everyone are tabulated for city blocks, while data from the census long-form sample (about 1 in 6 households) are tabulated for larger geographic areas. Statistical reliability is another reason besides confidentiality protection to limit the geographic detail of sample tabulations.

[13] Data swapping, which was first used in the 1990 census, replaced the previously used technique of cell suppression, in which cell values that were smaller than a specified threshold were blanked out. The suppression method made the data harder for users to work with (see Gates, 2000).

cestry to broad categories. Because of increasing concern about the ability to link census and survey microdata files with other data available through the Internet, the Bureau scaled back the data content somewhat on the 2000 census PUMS files in comparison with the 1990 census files.

Some microdata files of individual records are viewed as too sensitive and too easily re-identifiable to release in the form of a PUMS. For such data, the Census Bureau provides access to researchers who are sworn in as special census agents. For years such access could only be obtained by researchers who came to the Bureau's headquarters at Suitland, Maryland, to perform their analyses. In the past decade, the Bureau has begun a program of establishing secure research data centers at major universities, at which researchers may use data files that are not otherwise available for outside use (see Dunne, 2001). At present, there are six such centers: the Bureau's Boston Regional Office, Carnegie Mellon University, Duke University, the University of Michigan, and jointly managed sites at the Berkeley and Los Angeles campuses of the University of California.

Other Statistical Agencies

Other federal statistical agencies use similar methods to those of the Census Bureau to protect data during the stages of collection, processing, and storage and to minimize disclosure risks for data products that are made publicly available (see Federal Committee on Statistical Methodology, 1994). An additional source of concern for these agencies about disclosure risks during data collection and processing arises from the use of private contractors to conduct many of the household surveys they sponsor. (The Census Bureau uses its own staff for data collection for its surveys and those it conducts under contract to other agencies.) When contractors are used, agencies must carefully review the confidentiality protection procedures at contractors' sites.

For researcher access to sensitive data that are at risk for re-identification, some statistical agencies use licensing agreements. For example, NCES has statutory authority to sign licensing agreements that permit researchers to use microdata at their own institutions under specified restrictions (e.g., not sharing the data outside the research group, returning or destroying all copies of the microdata at the end of the project, etc.) The agreements must be signed by the researcher's institution, and they contain penalties for noncompliance. Other agencies use licensing agreements as well. Sometimes agencies audit data users' protection policies on a random or scheduled basis. (See

Seastrom, 2001, for a review of current licensing practices and requirements by federal statistical and program agencies.)

Finally, statistical agencies are investigating the use of new techniques for statistically perturbing sensitive microdata so that it may be possible to make them available in public-use form. Such methods include data swapping with additive noise and creating a synthetic data set through statistical modeling. Determining the net utility of such data sets requires estimating an index of information loss and one of disclosure risk and judging when there is an acceptable balance between the two (see discussion in Appendix E).

THE ROLE OF RESEARCHERS, IRBS, OHRP, AND FUNDING AGENCIES IN PROTECTING CONFIDENTIALITY

The Common Rule requires IRBs to determine that research proposals have adequate plans to protect the confidentiality of data obtained from respondents and to protect their privacy. Such protection is supported by the ethical principles in the Belmont Report. Yet we believe that IRBs, OHRP, and researchers may not be giving as much attention to issues of confidentiality protection as warranted by the increasing risks of disclosure from advances in technology and the volume and richness of available data. We believe it is critical that federal funding agencies support continued research on methods for confidentiality protection.

> **Recommendation 5.1:** Because of increased risks of identification of individual research participants with new methods of data collection and dissemination, the human research participant protection system should continually seek to develop and implement state-of-the-art disclosure protection practices and methods. Toward this goal:
>
> - researchers should explicitly describe procedures to protect the confidentiality of the data to be collected in protocols they submit to IRBs;
>
> - IRBs should pay close attention to the adequacy of proposed procedures for protecting confidentiality;
>
> - federal funding agencies should support research on techniques to protect the confidentiality of SBES data that are made available for research use; and
>
> - the Office for Human Research Protections should regularly promulgate good practices in analyzing disclosure risks and limiting those risks.

Researchers have an obligation to provide sufficiently detailed information in their proposal on plans for confidentiality protection so that an IRB can make an informed judgment about the adequacy of those plans. It is not enough to say that confidentiality will be protected—the methods and procedures for doing so at each stage of the research project must be detailed. Similarly, IRBs have an obligation to carefully review proposed plans for confidentiality protection and to evaluate them against recognized good practices that are applicable for the type of research proposed.

Federal funding agencies are increasingly interested in leveraging the dollars they invest in data collection under research grants and hence are requiring investigators to share data. Consequently, they have an interest in and, we believe, an obligation to support research on ways to analyze the risk of disclosure and on new methods for confidentiality protection that minimize disclosure risk and maximize the usefulness of shared data for secondary analysis. Such agencies could also partner with academic statisticians to disseminate information to researchers and IRBs about statistically based methods for disclosure risk analysis and risk minimization.

OHRP has a leadership responsibility for guidance on issues of human research participant protection. Because it is woefully inefficient for every IRB—many of which are overburdened—to take individual responsibility for staying abreast of threats to and state-of-the-art ways for protection of confidentiality, OHRP should regularly assemble and disseminate information on good practices for analyzing disclosure risk and minimizing that risk at every stage of a research project— from data collection to dissemination of results and sharing of data for secondary analysis. OHRP should also assemble and publish information on the confidentiality and data access guidelines of federal and state agencies with responsibility for administrative records that are of potential use for research. Such information would help researchers navigate the maze of varying agency policies and would also help IRBs evaluate research that proposes to use such data.[14]

In increasing their attention to confidentiality issues, we do not intend that IRBs (or OHRP) should add bureaucratic impediments to SBES research or waste scarce time and resources in activities that duplicate other efforts. We make four points in this regard and address

---

[14]The National Human Research Protections Advisory Committee adopted a similar recommendation at its April 29-30, 2002, meeting. The recommendation also urged OHRP to identify federal statutes and regulations that provide confidentiality protection, identify issues or gaps, and develop proposals to address these gaps through "a consensus process involving the scientific and legal communities" (see http://www.ohrp.osophs.dhhs.gov/nhrpac/documents [4/10/03]).

a fifth point in the next section:

(1) confidentiality protection should be appropriate to disclosure risk and the sensitivity of the data;

(2) adequacy of confidentiality protection should be assessed for each stage of a project involving original data collection—from recruitment to dissemination and archiving;

(3) IRBs should look to other bodies for guidance on good practices for confidentiality protection;

(4) informed consent processes and documentation should address the extent and nature of confidentiality protection; and

(5) IRBs should, as standard procedure, exempt from review studies that propose to use publicly available microdata files from sources that follow good protection practices and obtain informed consent from participants (see "A Confidentiality Protection System for Public-Use Microdata," below).

Confidentiality Protection Appropriate to Disclosure Risk

We have been stressing the risks of disclosure; however, there are many projects for which confidentiality protection is unnecessary or irrelevant or the needed protections can be very limited. Observational studies of anonymous individuals in public settings (e.g., shoppers at a store who are not approached directly by the investigator and are not photographed or videotaped) need no confidentiality protection at all. Oral history studies in which public officials are interviewed about their public activities may require only limited protection, such as respecting the right of the respondent to refuse to answer a particular question or putting an agreed-upon time restriction on the availability of the full oral history. Small laboratory experiments on stimulus-response behaviors may adequately protect confidentiality simply by not recording names or other identifiers of participants. Investigators in some participant observation studies may seek the consent of participants to include them individually in the published findings, likely with the use of pseudonyms, although participants must understand that pseudonyms will not necessarily protect them from being identified.

The larger point of all these examples is that for confidentiality protection, as in many other aspects of human research participant protection, there is no single approach that is appropriate for all studies. The risks of disclosure and the need for confidentiality protection should be analyzed for each type of project and confidentiality protections made more or less stringent as appropriate.

Guidance that OHRP develops on analyzing disclosure risk and implementing appropriate confidentiality protections should include examples not only of studies that require stringent confidentiality protection measures, but also of studies for which minimal or no confidentiality protection is needed.

Protection for Every Stage of Research

For projects that involve original data collection, IRBs will need to check that appropriate confidentiality protection procedures are proposed for each project stage, as applicable:

- recruitment of participants—protection practices will vary depending on the method of recruitment (e.g., sending a letter that contains specific information about the prospective participant requires more attention to confidentiality protection than does a random-digit telephone dialing procedure);

- training of research staff, including interviewers, computer processing staff, analysts, and archivists, in confidentiality protection practices;

- collection of data from participants—protection practices will vary depending on whether collection is on paper, by CATI, by CAPI, on the web, or by other techniques, and who is being asked for information (e.g., some studies of families allow individual members to enter their own responses into a computer in such a way that neither other family members nor the interviewer are privy to the responses);

- transfer of data to the research organization, whether by regular mail, e-mail, express mail, or other means;

- data processing (including data entry and editing);

- data linkage (including matching with administrative records or appending neighborhood characteristics);

- data analysis;

- publication of quantitative or qualitative results;

- storage of data for further analysis by the investigator or for recontacting participants to obtain additional data or both; and

- dissemination of quantitative and qualitative microdata for secondary analysis by other researchers.

For qualitative research, Johnson (1982) has developed advice on "ethical proofreading" of field reports prior to publication so that even if participants and their communities are identified, the harm to them is minimized. Her guidelines include such steps as reviewing language to make it descriptive rather than judgmental, providing context for unflattering descriptions, asking some of the participants to read the manuscript for accuracy and provide feedback, and asking colleagues to read the manuscript critically for ethical concerns.

Use of Authoritative Guidance

Until OHRP begins to promulgate good practices for confidentiality protection for different stages and types of projects, IRBs should seek out sources of guidance from reputable sources rather than developing standards for review of projects on their own. For example, many professional associations have developed and published good practices for confidentiality protection for studies in their discipline (see, e.g., Oral History Association Evaluation Guidelines; available at http://www.dickinson.edu/oha [4/10/03]). Major survey organizations also have principles and practices for confidentiality protection (e.g., see Institute for Social Research, 1999). For protection strategies for data that are to be published or shared with other researchers, see the paper we commissioned by George Duncan in Appendix E. See also the following resources: Czajka and Kasprzyk, 2002; relevant chapters in Doyle et al. (2001); Statistical Working Paper 22 (Federal Committee on Statistical Methodology, 1994); guidance from the ICPSR, available at http://www.icpsr.umich.edu/ACCESS [4/10/03]; and links to information resources provided by the American Statistical Association Committee on Privacy and Confidentiality, available at http://www.amstat.org/comm/cmtepc [4/10/03].

Confidentiality Protection and Informed Consent

In reviewing research that involves original data collection, IRBs need to consider the adequacy of the information about confidentiality protection that is provided to participants through the informed consent process (see also Chapter 4). For example, participants should be informed that the data will be made available for research purposes in a form that protects against the risk of re-identification. If identifiers such as social security numbers are requested to permit linkages with administrative records, respondents should be informed about steps that will be taken to prevent misuse of such identifiers and records and whether and when identifiers will be destroyed. The consent process should also make clear that confidentiality protection is never ironclad; rather, disclosure risks are minimized to the extent possible.

For research on illegal behavior (e.g., drug abuse) or sensitive topics (e.g., alcoholism, sexual abuse, or domestic violence), it is vitally important that adequate measures are in place to protect the privacy and confidentiality of research participants. Serious consequences may result if there is an intentional or inadvertent breach of confidentiality (including social stigmatization, discrimination, loss of employment, emotional harm, civil or criminal liability, and, in some cases, physical injury). Investigators must ensure that the informed consent discussion delineates carefully the procedures for protecting confidentiality, which may include waiving written consent or obtaining a certificate of confidentiality to prevent data from being used in court. In addition, investigators must address the possibility that they may have to report such behaviors as child abuse to authorities.

A CONFIDENTIALITY PROTECTION SYSTEM FOR PUBLIC-USE MICRODATA

Recommendation 5.2: To facilitate secondary analysis of public-use microdata files, the Office for Human Research Protections, working with appropriate federal agencies and interagency groups, should establish a new confidentiality protection system for these data. The new system should build upon existing and new data archives and statistical agencies.

Recommendation 5.3: Participating archives in the new public-use microdata protection system should certify to researchers whether data sets obtained from such an archive are sufficiently protected against disclosure to be acceptable for secondary analysis. IRBs should exempt such secondary analysis from review on the basis of the certification provided.

We argue that IRB review of secondary analysis with public-use microdata is unnecessary and a misuse of scarce time and resources (see Chapter 6). If the data in a file have been processed to minimize the risk of re-identifying a respondent by using widely recognized good practices for confidentiality protection, then the research is eligible for exemption under the Common Rule (see Box 1-1 in Chapter 1). The issue is how an IRB can be satisfied that a particular public-use microdata file has been processed using good practices for confidentiality protection. To address this concern, we propose that OHRP work with statistical agencies, data archives, and appropriate interagency groups to develop a new system for confidentiality protection and certification for public-use microdata. Such a system would permit IRBs to exempt secondary analysis with such data from review as a matter of standard practice.[15]

We have described how federal statistical agencies are in the forefront of efforts to protect the confidentiality of their data. When such agencies release a public-use microdata set (or summary file for small geographic areas), one can be assured that they have followed good practices for confidentiality protection. One can also be assured that they have addressed such aspects of human participant protection as informed consent and minimization of respondent burden because of the requirement that all data collections by federal agencies be cleared by OMB under the Paperwork Reduction Act (some agencies also have an IRB).

We recommend that OHRP work with the Interagency Council on Statistical Policy, which includes 14 statistical agencies and is chaired by the chief statistician in OMB, to develop a certificate that accompanies release of public-use data sets from these agencies on the web or in other media. Such a certificate would attest that the public-use file reflects good practice for confidentiality protection and that the data were collected with appropriate concern for informed consent and other protection issues. With such a certificate, the IRB would exempt from further review any analysis that proposes to use only the data from the certified file.

Public-use microdata are also made available from federal program agencies and from private archives and research organizations. To extend the certification system, OHRP should work with the Interagency Council on Federal Statistics, other public and private data producers, and data archives to develop something like the assurance program that ORHP uses to authorize IRB operations at research institutions under the Common Rule. Under such a confidentiality assurance program, an archive such as ICPSR would document the procedures it uses to protect confidentiality for data sets that researchers deposit with the archive for redistribution to secondary analysts. Once its procedures are approved, the archive would be able to certify that

---

[15]Summary data that are provided for small geographic areas or population groups could also be certified; we recommend in Chapter 6 that analysis with such data not be brought to IRBs, as the aggregate data no longer represent human subjects under the regulations.

the data files it distributes are appropriately processed for confidentiality protection.[16] Similarly, other data producers, such as federal program agencies and private research organizations, could obtain organization-wide certification for their public-use files, or an organization could obtain certification on a case-by-case basis if it rarely develops public-use data.

A program of assurance for confidentiality protection procedures and certification of data files for secondary analysis will necessitate that participants in the program—OHRP, federal statistical agencies, other data producers, and archives—keep abreast of disclosure risks and state-of-the art protection procedures. Continued vigilance, together with sustained investment in disclosure risk analysis and confidentiality protection methods, will be necessary to assure IRBs, researchers, and participants that adequate protections are in place.

CONCLUDING NOTE: MINIMAL DISCLOSURE RISK IS NOT ZERO RISK

At present, there is considerable tension between the SBES research community and data producers, particularly federal statistical agencies, regarding what and how much microdata can be made available for public use. Statistical agencies, in some researchers' views, are striving for zero disclosure risk, which is not possible, and are unnecessarily restricting the availability of data that were collected with public funds and intended for public use. Researchers, in the view of many statistical agencies, are underestimating the disclosure risks and are not sufficiently cognizant of the legal constraints and penalties under which statistical agencies operate.

We cannot resolve the tensions between these views. Several studies of the Committee on National Statistics have addressed confidentiality issues (National Research Council, 1993, 2000a), and a study is currently under way to address specifically the balance of benefits and costs of data access versus disclosure risk. Some solutions will likely require congressional action, such as legislation that would enable more statistical agencies to use licensing agreements that make the researcher, as well as the statistical agency, responsible for any breach of confidentiality. Other solutions will require accommodation of views. For example, researchers may have to be more accepting of conducting secondary analysis at secure data centers, while the spon-

---

[16]ICPSR has informed researchers in member institutions on how to obtain information from its website on ICPSR confidentiality protection procedures to accompany protocol submissions to IRBs (Erik Austin, 2002, personal communication).

sors of data centers may need to interpret more broadly the kinds of studies that are acceptable to be conducted at a secure center. Currently, for example, the Census Bureau approves studies that will help the Census Bureau (e.g., studies of missing data), which seems too narrow a criterion given that the mission of the Bureau is to provide information for public use.

Given the leverage that secondary data analysis provides for the advancement of knowledge in the social, behavioral, and economic sciences, it is clearly important for researchers, data producers, and data archives to work cooperatively to maximize that leverage while appropriately protecting the respondents who supplied the data. IRBs can contribute to such a cooperative effort by encouraging investigators who collect original data to deposit it with archives that will make the data available to others in a form that minimizes the risks of breach of confidentiality.

— 6 —
Enhancing the Effectiveness of Review: Minimal-Risk Research

I N THIS CHAPTER we address the operation of institutional review boards (IRBs) with respect to review procedures for minimal-risk social, behavioral, and economic sciences (SBES) research. The available evidence, although limited (see Chapter 2), supports a conclusion that IRBs often decide not to use the flexibility in the Common Rule that allows them to exempt some types of SBES (and biomedical) research from review or to review other kinds of minimal-risk SBES (and biomedical) research with an expedited procedure. Some IRBs do not exempt or expedite any research, even when they agree that it is minimal risk, and others do not do so for certain categories of eligible research (see Boxes 1-1 and 1-2 in Chapter 1 for categories of exempt research and categories of research that can be expedited, respectively).

Furthermore, although there is only anecdotal evidence on this point, IRBs in some instances may overestimate the risks of harm to participants in SBES (and biomedical) research. In such cases, they may use a time-consuming full board review when it is not needed and perhaps request changes in research design that compromise the scientific validity of the study without necessarily increasing protection for participants. At the other extreme, a few IRBs may underestimate risk given that they never or hardly ever conduct full board reviews.

The primary goal of the IRB system is to conduct reviews that protect participants to the extent possible. The Common Rule sets out alternative procedures for minimal-risk research that are adequate for protection and can free up scarce resources of IRBs and investigators to devote more attention to higher risk research when needed and to other activities that are important for participant protection, such as the development of improved guidance and training for IRB members, researchers, and research institution officials (see Chapter 7). The challenge in the difficult environment in which IRBs operate today is to encourage them to make appropriate use of the flexibility in the regulations.

We begin this chapter by explaining why we think that the Office for Human Research Protections (OHRP) in the U.S. Department of Health and Human Services (DHHS) should develop detailed guidance for IRBs on the designation and treatment of minimal-risk SBES (and biomedical) research. We then discuss guidance that addresses each of the provisions of the Common Rule that are designed to help the effectiveness of the review process for minimal-risk research, beginning with definitions of "research" and "human subject" and proceeding to criteria for exempting research and for using an expedited review procedure. We also consider guidance on effective procedures for ongoing review of minimal-risk research.

The last two sections of the chapter discuss the need for better data of two types. First, there is a need for data about perceived risks and actual harms encountered by participants in SBES research to better inform classification of research protocols. Second, there is a need for data on the operation of the IRB system. At present there is no way to track changes in the central tendency or variability of IRB operations, such as how many IRBs are using expedited review procedures appropriately.

Our focus on IRBs in this chapter is not meant to imply that they are the only source of concern in the operation of the U.S. human research participant protection system. All actors in the system can and should do more to reinforce commitment to participant protection. In particular, we believe it is incumbent upon the research community to do more than raise concerns about IRB review procedures. Researchers need to be proactive in developing knowledge that will help inform assessments by IRBs and investigators of harm and risks for various types of research and help determine appropriate procedures for participant protection in research protocols.

GUIDANCE ON THE REVIEW PROCESS

OHRP should develop detailed guidance for IRBs on using the minimal-risk provisions of the Common Rule for four reasons. First, no IRB can be expected to have the expertise among its members to understand the particular issues and requirements for all of the kinds of research it is likely to encounter, even if the IRB is charged to review research that is clearly in one domain (see Chapter 7). Enlisting other researchers with appropriate expertise to help review individual protocols can help (the 1995 Bell survey found this to be the practice of many high-volume IRBs—see Chapter 2). But we also see a role for OHRP to assist IRBs to handle a diverse workload by providing specific guidance for a variety of research topics, methods, and study populations on such matters as when it is appropriate to exempt a protocol or to use the expedited review process.

Second, there is evidence that greater specificity of guidance leads to a greater likelihood that IRBs will in fact follow the guidance. For example, the 1995 Bell survey found that IRBs were most likely to use an expedited procedure, as a matter of standard practice, to review minimal-risk protocols that involved such methods as collecting nail and hair clippings and scraping dental plaque. Taking high-volume and low-volume IRBs together, 75 percent had such a practice for nail and hair clippings, and 66 percent had such a practice for dental samples. In contrast, IRBs were least likely, as a matter of standard practice, to use an expedited procedure to review minimal-risk protocols that involved a drug or device—only 26 percent had such a practice (Bell, Whiton, and Connelly, 1998:30). We believe one likely reason for this difference is that, in the regulatory listing of research eligible for expedited review (provided the research is minimal risk), nail and hair clippings and dental plaque are very specific items, while drugs and devices cover a wide range of (unspecified) items (see Box 1-2). Hence, we expect that greater specificity regarding minimal-risk research involving surveys, interviews, observations, and other typical SBES methods could encourage more IRBs to use an expedited review procedure for such research than do so at present.

A third reason for providing more detailed guidance on such matters as exemptions and expedited review is that greater use of the guidance could reduce variability in IRB procedures (see Chapter 2). Historically, reasons for the decentralized structure of the IRB system, in which there are many individual IRBs, each with the power to determine its own procedures (so long as they are at least as rigorous as the Common Rule), were to facilitate local community input to the research review process and promote responsiveness to community norms and practices. Local input is important, particularly when research involves community residents as participants. However, we do not believe it likely that differing community norms explain the current wide variation in such practices as exemption and expedited review policies. Reducing variability in IRB practice would not, in our view, undermine the principle of community input and would greatly assist many researchers who move from one institution to another and the growing number who are involved in multisite studies.

Finally, at present, only limited, scattered guidance is available to IRBs on ways to use review procedures that are appropriate to the level of risk. Some research sponsors have provided specific guidance to IRBs. For example, the Centers for Disease Control and Prevention has a document that provides examples of public health investigations that do and do not constitute "research" (Centers for Disease Control and Prevention, 1999). The recently issued National Science Foundation (2002) guidance also has some useful examples, but this document is aimed at researchers primarily and at IRBs only indirectly. The *IRB Guidebook* (Office for Human Research Protections, 1993, available http://ohrp.osophs.dhhs.gov/irb/irb_guidebook.htm [4/10/03]) provides useful general guidance but gives very few specific examples. We believe that for IRBs to move toward greater use of Common Rule provisions that are appropriate for minimal-risk research in the current climate of intense scrutiny of IRB actions will require detailed authoritative federal guidance from a body with broad oversight powers. In our view the appropriate body to prepare and implement such guidance is OHRP.

GUIDANCE FOR INITIAL REVIEW

Recommendation 6.1:   To promote review appropriately tailored to risk, the Office for Human Research Protections should develop detailed guidance for IRBs and researchers (with clear examples for a variety of methods) on what kinds of social, behavioral, and economic sciences (SBES) research protocols qualify as "research" with "human subjects." OHRP should also develop detailed guidance, including examples, regarding SBES research that IRBs are strongly encouraged to exempt from review and research that IRBs are strongly encouraged to review with an expedited procedure.

Below we consider some of the issues affecting each decision in the process whereby IRBs determine the type of review to afford a newly proposed protocol and offer some examples of possible guidance for different kinds of SBES research. We look forward to discussion in the research and IRB communities to inform action by OHRP on guidance that includes a range of useful examples. OHRP should issue such guidance as soon as possible and add to it and modify it as appropriate in future years (see Chapter 7 for discussion of a process for developing OHRP guidance).

What Is Research?

The Common Rule (45 *CFR* 46.102e) defines "research" as "a systematic investigation, including research development, testing and evaluation, designed to develop or contribute to generalizable knowledge." Research activities that are part of demonstration and service programs are included, as well as stand-alone research. The Common Rule also states that research may be conducted by a professional or student investigator.

This definition has two aspects. First, it refers to a set of activities that involve a human participant in a research experience, that is, a "systematic investigation." Second, it states that, regardless of the nature of the participant experience, that experience is research only if it is intended to produce or contribute to generalizable knowledge. Because the line between a systematic investigation and exploratory activities preparatory to research may be unclear, and because the same investigation may or may not be research depending on the intent, disagreements may well arise in specifying what is and is not research.

We agree that the definition of research should include such developmental activities as pretests and structured focus groups when they are integral to the design and planning of a larger research project (e.g., one or more focus groups to test the psychological sensitivity of survey questions). However, not all planning activities constitute research. Purely exploratory activities, such as contacting representatives of employers to determine their policies for cooperating with surveys of their employees, or asking a few colleagues to react to a questionnaire format, do not, in our view, fit any definition of reviewable research.

A type of activity involving a "systematic investigation" that may or not be "research" under the Common Rule is research with human participants conducted by undergraduate students in SBES courses in political science, economics, psychology, and other subjects. For example, students in a course on American politics may be asked to design and conduct a small survey of community residents or fellow students on a current public policy issue. If the sole intent of the exercise is to teach survey construction and not produce generalizable knowledge, the exercise, by the definition above, is not research. Yet if the exact same survey is conducted to gain generalizable knowledge related to the public policy issue (as might be the case in a senior thesis), then the survey qualifies as research. Finally, if the data are collected simply to teach the method but, once obtained, are deemed of sufficient quality and interest to submit for publication, then the same survey would change from not being research to being research.

Not surprisingly, IRBs differ in their approaches to reviewing undergraduate research. From our review of IRB websites for 47 major research institutions, it appears that 75 percent of these IRBs require that all undergraduate research projects be submitted for IRB review.

Six percent do not have such a requirement, and the remaining 20 percent require review of selected projects (e.g., senior theses) or of the instructor's plans for student research for the course as a whole (see Appendix D).

We agree with the importance of instilling an awareness of human research participant protection issues among undergraduates, and there is also always the possibility that an undergraduate may design a project that poses a serious concern. For example, a student might design a laboratory experiment to test differences in speed of perception between sober and inebriated fraternity members. Yet as our website review reveals, there are likely more effective and efficient procedures for needed ethical review of undergraduate research projects—whether or not they meet the strict definition of "research"—than individual project review by IRBs. For example, the IRB could review instructors' course plans for student projects and have the instructor, who is the responsible principal investigator, provide the individual project reviews (see Appendix D). Alternatively, an IRB under its mandate to educate the research community about ethical requirements for human participant protection could provide a school or department with guidelines for review of student projects which should be of no more than minimal risk. It would then be the relevant department's obligation to conduct the appropriate review in accordance with the guidelines. Projects falling outside of the guidelines would be subject to IRB review.

Who Are Human Subjects?

The Common Rule (45 *CFR* 46.102f) defines a "human subject" (participant) as a "living individual about whom an investigator (whether professional or student) conducting research obtains (1) data through intervention or interaction with the individual, or (2) identifiable private information." The regulations go on to stipulate that:

> *Intervention* includes both physical procedures by which data are generated (for example, venipuncture) and manipulations of the subject's environment for research purposes. *Interaction* includes communication or contact between the investigator and subject. *Private information* includes information about behavior that occurs in a context in which an individual can reasonably expect that no observation or recording is taking place, and information which has been provided for specific purposes... which the individual can reasonably expect will not be made public (for example, a

medical record). Private information must be individually identifiable... in order for obtaining the information to constitute research involving human subjects. [italics in the original]

Deciding when research involves human subjects or participants under the regulations is not always straightforward. In some instances, the sought-after information might not seem to be "about" the individual—for example, asking survey respondents about world events or governmental programs. Yet such responses can inform the researcher about attributes of the individual, such as the person's political stance, attitudes toward authority, or awareness of current events, and, hence, the survey does constitute research with human participants.

In other instances, humans may contribute importantly to research projects in ways that do not and should not involve them as participants under the regulations. One example is when humans are asked to provide data about organizations but are not themselves an object of study. A second example is when research is conducted using data that were originally obtained from human participants but that are provided in an aggregate or tabular form from which information about individuals cannot be recovered. We elaborate further on these two examples below.

Research on Organizations

SBES researchers study not only individuals and groups, but also organizations, such as businesses and governments. For example, business economists may seek to relate firm size, measured by number of employees or value of products sold, to measures of innovation, such as spending on research and development or patent applications. Public finance researchers may seek to relate spending by local governments to changes in the business cycle. For such projects, it is usually necessary to ask one or more individuals to provide needed information about the organization from relevant organizational records (e.g., accounting or personnel files). When individuals are interviewed in their capacity as knowledgeable agents of an organization and not as the direct objects of inquiry, and no attempt is made to include any of their characteristics in the study analysis, then the analysis is not likely research with "human subjects."

Organization research often raises issues that are similar to research with human participants in regard to protecting confidentiality (of organizations in this case) and providing sufficient information for an informed decision by the organization about participation. However, the IRB system is designed to address such issues for human participants and not organizational entities, which are presumed to be able to protect their own interests. Consequently, IRBs should not use their scarce resources to review research that involves humans solely as agents of an organization.

Secondary Analysis with Aggregate Data

SBES researchers in many fields conduct secondary analyses of aggregate data collected by another researcher, research organization, or statistical agency. An example of aggregate secondary analysis is research conducted by geographers and sociologists on migration flows between states, counties, and cities from tabulations of the U.S. decennial census long-form sample. Such analysis is at the level of groups, not individuals. As we describe in Chapter 5, prior to publication the Bureau takes steps to ensure that manipulation of the individual cells of a tabulation will be extremely unlikely to reveal the identity of individual respondents.

When aggregate data are obtained for geographic areas or population groups from a source that does not have a known track record of confidentiality protection (e.g., an individual investigator), then researchers may need to seek IRB review before proceeding to use such data in analysis. However, when analysis is planned of publicly available tabular data from such sources as the Census Bureau and other federal statistical agencies, there should be no need for researchers to seek IRB review. The research does not involve individual human participants and no issues of protection of humans arise.

Exemption

After determining that a protocol clearly covers human subjects, the next stage in the IRB decision process is to determine if the protocol qualifies for and should be exempted from IRB review. There are six categories of research activities that are eligible for exemption, as specified in the Common Rule (see Box 1-1). The Common Rule does not specify that exempted research must be of minimal risk; however, the categories are clearly focused on minimal-risk research or, in the case of public service evaluation programs, on research that is under the direct review of department or agency heads.

As of 1995 (from the Bell survey; see Chapter 2), sizable proportions of IRBs were not exempting eligible research in one or more of these categories as a matter of standard practice; 35 percent reported never giving an exemption. Most commonly, IRBs were reviewing research that was eligible for exemption with an expedited procedure (Bell, Whiton, and Connelly, 1998:28). Our review of IRB websites of 47 major research universities in late 2002 showed that most will consider granting an exemption, although a few (9 percent) will not.[1] We acknowledge that it can be difficult to judge the appropriateness of exempting a particular research project from review even when it appears to be eligible for exemption. For example, the consumer telephone survey example in Box 2-5 (in Chapter 2) could qualify for exemption under category (2): it does not collect identifying or potentially damaging information. Yet an IRB might want to review such a survey with an expedited procedure to satisfy itself that the interviewer script provides adequate information to obtain respondents' informed consent.

Whether an IRB decides to exempt such research might well depend on how effective the IRB believes it has been in providing guidance and training on ethical research practices to investigators at its institution. We argue that the development by OHRP of guidance for IRBs on exemption practices could encourage higher percentages of IRBs to exempt eligible research and thereby conserve scarce resources to use on protocols that merit IRB attention and on such activities as development of improved training. IRBs may also want to systematically evaluate their policies and procedures for determining exempt status, such as rotating the duty among members to assess the consistency of members' implementation of the IRB's exemption policy, or sampling a subset of exempted protocols to assess the decisions made.

Exemption relieves an investigator from the obligation to undergo IRB review; however, it does not relieve him or her of the duty to adhere to the ethical principles of the Belmont report. Investigators are not empowered to make decisions themselves on whether a research project with human participants qualifies for exemption; they are required to submit the appropriate information to request an exemption. However, some IRBs (33 percent in the 1995 survey) routinely accept an investigator's declaration of eligibility for exemption;[2] other IRBs (67 percent in the 1995 survey) independently determine exempt status (Bell, Whiton, and Connelly, 1998:28-29).

Below we discuss two kinds of research that we believe IRBs should exempt from review as a matter of standard practice: observational studies of public behavior when the investigator has no contact with participants and secondary analyses of public-use data for individuals

---

[1] Of this 9 percent, half stated that the IRB does not grant exemptions; the other half made no mention of and provided no way to seek exemption.

[2] It is not clear from Bell, Whiton, and Connelly (1998) whether the chair or the IRB administrator routinely determines exempt status.

(microdata), when the data are obtained from suppliers, such as federal statistical agencies and data archives, that regularly follow good practices to minimize the risk of identification of individuals. Given that confidentiality of participants is protected, these types of research are *prima facie* of minimal risk because they do not involve any intervention or interaction that could pose a risk to an individual participant.

Observational Studies of Public Behavior

Many studies by social psychologists and cultural anthropologists involve observing the behavior of people in public places going about their ordinary business who are not aware of the observation. In such situations people should be able to expect that they will remain anonymous, but they should not have an expectation of privacy: they are in a public place (e.g., on a street, in a building lobby, in a government building, in a public park), which may be observed not only by researchers, but also by journalists, civil or criminal investigators, and casual loiterers or passers-by.

This type of observational research generally qualifies for exemption from IRB review under category (2) of the Common Rule listing of exempt research categories (see Box 1-1). For example, the observational study of pedestrians crossing a street (described in Box 2-3) meets the requirements for exemption: the investigator takes notes of what occurs but makes no attempt to interact with the observed pedestrians. The investigator's notes may record such information as the sex, race, age, and type of dress of pedestrians (as determined by observation), but this information would scarcely permit identification of individual participants if it became known to others, and disclosure would not likely place the participants at risk of any legal, economic, or social harm.

If the investigator were to photograph or videotape the participants as an aid to analysis, then the photographs or videos could possibly identify individuals. Yet photographic records should not automatically exclude the possibility of exemption. Exemption can be granted under category (2) when respondents are identifiable so long as disclosure outside the research team could not "reasonably place the subjects at risk of criminal or civil liability or be damaging to the subjects' financial standing, employability, or reputation."

Secondary Analysis of Public-Use Microdata

Much SBES research requires microdata containing numerous variables on individual people or households to use in multivariate analysis. For understanding such behaviors as voting, labor force participation, and welfare program participation, aggregate data are generally not richly detailed enough to enable estimation of relationships. The problem for the researcher is that collection of microdata from large samples involves substantial up-front costs. The researcher can avoid these costs if he or she can gain access to microdata that someone else has already collected.

Beginning in the late 1960s, when computers first became widely used for data analysis, the number and richness of available public-use microdata files from federal statistical agencies and other sources has grown enormously (see Chapter 5). These files, which the source agency processes to minimize the risk that individual respondents can be re-identified, add significantly to the infrastructure for cost-effective research in many SBES fields.

Technological developments—in particular, the widespread availability of administrative records and other data on the Internet that can possibly be linked with microdata sets, especially when the microdata are richly detailed (e.g., longitudinal surveys)—are making it harder to construct public-use microdata files that minimize disclosure risks. These developments may make it necessary in some instances to scale back the richness of the data that are included on a public-use file or to use statistical techniques to perturb the data in ways that may make it harder to estimate relationships. We discuss these challenges and potential solutions to them in Chapter 5. Here we argue that individual IRBs should not second-guess the confidentiality protection measures that have been implemented to minimize the risk of disclosure from a public-use microdata file. Instead, IRBs should accept proper certification from researchers that the file they propose to use is a public-use microdata file as we have defined it and, therefore, that the proposed research (assuming it does not use other, identifiable data) is exempt from further IRB review under category (4) of the regulations.

We describe the kind of certification procedures we propose in Chapter 5. They are similar to those recommended by the National Human Research Protections Advisory Committee, which acted in response to a presentation from its Social and Behavioral Science Working Group at a committee meeting, January 28-29, 2002 (see http://ohrp.osophs. dhhs.gov/nhrpac/documents/dataltr.pdf [4/10/03]). Some might argue (as the working group did initially) that the use of certified public-use microdata files is non-human-subjects research on the grounds that human participants are not identifiable. We argue instead that such research qualifies as research with human participants because the risk of individual identification on public-use microdata files, while close to zero, is never zero. Therefore, it is appropriate for IRBs to be informed of proposed studies of public-use microdata files from a certified source, but they should exempt such studies as a matter of standard practice because participants have already been protected to the extent possible.

Expedited Review

If a human subjects research protocol does not meet the criteria for exemption, the next stage in the IRB decision process is to determine if it qualifies for expedited review. Expedited review means that the IRB chair or a subset of other members (or both) conducts the review, rather than the full board at a meeting. An expedited review is as comprehensive as full board review in the sense that the IRB chair (or subgroup) follows all of the applicable requirements of the Common Rule. However, it can be conducted in a much more expeditious manner with regard to elapsed time, and it saves on the time required for IRB business totaled over all IRB committee members (see Chapter 2).

The Common Rule authorizes publication in the *Federal Register* of a list of categories of research that, if the research is minimal risk, may be reviewed by an expedited procedure. The list may be amended, as appropriate, from time to time. The original list of such research was published in 1981, and the list was revised in 1998 (see Box 1-2). To qualify, the proposed research must be minimal risk, must fall into one of the approved categories, must incorporate steps to minimize the risk of breach of confidentiality in instances "where identification of the subjects and/or their responses would reasonably place them at risk of criminal or civil liability or be damaging to the subjects' financial standing, employability, insurability, reputation, or be stigmatizing" (see Box A-5 in Appendix A), and must not be classified.

We focus our attention on category (7), which relates most directly to SBES research. It covers "research on individual or group characteristics or behavior (including, but not limited to, research on perception, cognition, motivation, identity, language, communication, cultural beliefs or practices, and social behavior) or research employing survey, interview, oral history, focus group, program evaluation, human factors evaluation, or quality assurance methodologies." At the time of the 1995 Bell survey, fully 51 percent of IRBs, as a matter of standard practice, did *not* expedite any minimal-risk SBES research of the types included in category (7); 15 percent of IRBs did not expedite research in *any* category. Our review of IRB websites of 47 major research universities in 2002 showed that 13 percent did not expedite research in any of the eligible categories.

We present examples of laboratory behavioral research, interview research (specifically, oral histories), and survey research that we believe most likely qualify for expedited review and are candidates for OHRP to include in guidance to IRBs. The intent of OHRP guidance should be to provide a level of specificity comparable to that provided for biomedical research in categories (1) through (4) of the current list of research eligible for expedited review (see Box 1-2).

Laboratory Behavioral Research

Box 2-1 in Chapter 2 provides two examples of types of laboratory research that readily qualify for review with an expedited procedure. The first example is an experiment about economic decision making (e.g., reaching agreement on terms of exchange). A small number of participants are brought together and given precise, detailed instructions about how they are to interact and how they will be rewarded on the basis of their decisions and the decisions of other participants. They are informed they may leave at any time. Their decisions are recorded, and they are rewarded accordingly (in private, anonymously, and after the experiment). Reward amounts are small. No personal identifiers are kept. There seems to be no reason for full board review of such a study except when the proposed reward amount exceeds some threshold—which the IRB could specify—that could reasonably be viewed as inappropriate. Indeed, perhaps the only reason for the IRB to review such studies at all is that they commonly involve students, and the IRB may want to assure itself that students are not pressured to participate.

The second example is a social psychology experiment with deception. The research question is the extent to which people engage in ethnic stereotyping. A small number of participants are brought together and told that the purpose of the experiment is to determine how fast people can associate characteristics (e.g., good, bad) with lists of names (which differ in cues about ethnic origin). They are told they may leave at any time. Their results are recorded, and they are told at the conclusion of the experiment about its true purpose. No personal identifiers are kept from the experiment. Such experiments, even though they involve deception, do not require full board review. The stimulus (list of names) is not threatening, the only deception involves the purpose of the stimulus, and participants are fully debriefed. Moreover, the risk of breach of confidentiality is not an issue. In contrast, if the stimulus were threatening (e.g., aggressive behavior from investigators masquerading as participants), then the study would warrant full board review.

Interview Research

Oral histories constitute a type of research using unstructured or semistructured interviews that often qualify for review with an expedited procedure. Most such histories are designed to obtain information from an individual participant that the participant agrees will be released in full (or in part) at a specified time according to a legal agreement. The participant expects confidentiality only until the date when he or she has agreed that the interview may be made publicly available for historical research and other specified purposes (e.g., use in a television documentary). In the 1998 revision of the list of expeditable research, oral history was specifically included in the list of types of SBES research in category (7) that IRBs could review with an expedited procedure.[3]

The Oral History Association has developed extensive guidelines for the conduct of oral histories on such topics as informed consent, archiving and protection of confidentiality, legal releases governing when interviews may be made available, and many other aspects of ethical research using this method (see "Principles and Standards" and "Oral History Evaluation Guidelines" at http://www.dickinson.edu/oha [4/10/2003]). We believe that OHRP guidance could draw from these guidelines to encourage IRBs to develop a checklist that would allow for expedited review of many types of oral history as a matter of standard practice. The checklist could focus on the major sources of risks to human participants from oral histories—the risk of breach of confidentiality during the period before authorized release, the risk that some participants may not fully understand they have agreed that their personal histories eventually will be publicly available, and the risk that third parties who have not given consent may be adversely affected—and responses from investigators used accordingly to determine the type of review.

Survey Research

Many SBES surveys contain financially or psychologically sensitive content or raise other issues (e.g., third-party consent—see Chapter 4) that merit full board review by an IRB. However, many surveys could just as effectively be reviewed with an expedited procedure, qualifying for such review under category (7) of the current list of expeditable research (see Box 1-2).

---

[3] Oral history qualifies for exemption when it is conducted with elected or appointed public officials or candidates for public office.

Box 2-5 in Chapter 2 provides an example of a minimal-risk survey that qualifies for review with an expedited procedure (or even exemption from review). The survey is a short telephone survey of a sample of one adult in each of 1,000 households selected by random digit dialing. The content pertains to expectations about the state of the economy. Minimal content is collected about the household itself (e.g., number of members), and no identifiers are obtained.

Even longer and more complex surveys on such topics as attitudes and expectations could be effectively reviewed with an expedited procedure, if they do not retain identifiers and the questions being asked are not embarrassing or threatening. Examples are the University of Michigan Survey of Consumers and the Conference Board Survey of Consumer Attitudes and Buying Plans. The Michigan survey (see http://www.sca.isr.umich.edu [4/10/03]) interviews 500 people by telephone each month, drawing the sample from a list of household telephone numbers. Households are recontacted one more time 6 months after the original interview. The Conference Board survey (see http://www.conference-board.org [4/10/03]) sends a mail questionnaire to 5,000 people each month. The two surveys ask similar questions. The Michigan survey includes about 50 core questions. The majority of the questions are about the respondent's expectations on such topics as where the economy is heading and whether the household's income is likely to go up or down; a few questions ascertain basic demographic characteristics, including household size, number of children, and marital status, race, and education of the respondent. Were it not for the fact that the survey has a re-interview procedure (to produce stability in the estimates), which necessitates retaining telephone numbers, and is an ongoing survey that is subject to change, it could well be exempted from IRB review; in fact, it receives an annual expedited review (Rebecca McBee-Bonello, Survey Research Center, University of Michigan, January 28, 2003, personal communication).

CONTINUING REVIEW

Recommendation 6.2: Institutional review boards should use efficient procedures to review minor changes to minimal-risk research protocols that arise during the period of authorization. When appropriate, IRBs should approve protocols that allow researchers flexibility in making specific design decisions during the course of their research without the need to seek further review. (An example would be one of two forms of a question—both minimal risk—to be decided on the basis of a pretest.)

The Common Rule requires IRBs to "conduct continuing review of research covered by this policy at intervals appropriate to the degree of risk, but not less than once per year" (45 *CFR* 46.109c). Another provision (45 *CFR* 46.110b) permits IRBs to use an expedited procedure to review "minor changes in previously approved research during the period (of one year or less) for which approval is authorized."[4] Researcher concerns about these provisions are similar to those about the functioning of IRBs in general: that continuing review is perfunctory and not sufficient for human participant protection and that the requirement for IRB approval of changes is cumbersome and imposes needless delays, even when the changes are minor and the review is expedited. Below we address some issues and suggest some ways to facilitate review of minor changes to minimal-risk research.

In our discussion of common SBES research methods in Chapter 2, we note that one such method involves unstructured (or semistructured) interviews in which the content and scope of the questioning evolves over the course of the project and may differ for different participants. It would be highly disruptive to the research to bring every such change back to an IRB for approval. What is needed, instead, is for an investigator to provide the IRB with information at the time of initial review about contingencies that may arise. Perhaps the information provided could include illustrative interview paths that the investigator anticipates may be taken by different kinds of participants. If the research is minimal risk and the anticipated contingencies are also minimal risk, the IRB should approve the research without requiring the investigator to bring back every change in measurement for review. Instead, the investigator should be charged to come back to the IRB if the measurement evolves in unexpected ways that could raise the risks to participants.

A similar situation arises in a survey when the investigator is uncertain about which version of a question will better elicit the desired information and wants to conduct a pretest or focus group analysis before making a decision on how to word the question. Often, it will be reasonable for an IRB to require review of a pretest and a separate review of a final survey questionnaire. However, in the case of a minimal-risk survey in which the scope of the pretest is limited (e.g., a few questions) and both versions of the question(s) being tested are minimal risk, then the IRB could approve the protocol in one review operation. That is, a reasonable option would be to approve not only

---

[4]IRBs may also use an expedited review procedure to conduct annual reviews of research that is in the data analysis phase and of research that does not fall under categories (2) through (8)—see Box 1-2—but that the IRB has previously determined to be minimal risk.

the pretest, but also the survey, with the understanding that the survey may use either version of the question(s) that is supported by the pretest results.

Finally, IRBs should seek to learn from the practices of other IRBs that have particularly efficient methods for handling the review of changes to research protocols. For example, some IRBs allow researchers to inform the IRB of a change and, in turn, for the IRB to approve the change, entirely by electronic interchange. Electronic communication methods can undoubtedly facilitate IRB initial reviews as well, to reduce the time from submission of a protocol to a decision about the type of review and the time required in negotiation to produce an acceptable protocol.

DOCUMENTING RISKS AND HARMS

Recommendation 6.3: In order to build knowledge of research risks, OHRP and funding agencies should encourage researchers to build into their studies such steps as debriefing participants to learn about types, incidence, and magnitude of harm encountered in social, behavioral, and economic sciences research. Researchers should seek publication of their results.

As we discuss in Chapter 2, there is very little evidence on the distribution of harm experienced by participants in SBES research. Some types of SBES research, such as surveys, are believed by some investigators to have been conducted for decades with no serious physical or psychological harm to respondents. There are examples of studies in which psychological harms occurred (see Box 3-1 in Chapter 3), but whether the research caused long-lasting harm is not known. There are also examples of research that demeaned participants and did not respect their autonomy. Yet there is hardly any quantitative data on risks and harm of SBES research that could inform policy on human research participant protection or contribute to improved research practice.

We believe it is incumbent on SBES researchers, as part of sound, ethical research practice, to adopt procedures to debrief respondents (or samples of them) on their perceptions of risk and experiences of harm and to include the results in published accounts of the research. For example, survey respondents could be asked whether they felt coerced to participate, whether they found any questions to be troubling, whether any adverse reactions they experienced to questions were intense or expected to linger, and so on. Such debriefing could be part of the study itself or conducted as a follow-on. Post-study interviews or debriefings would need careful design to minimize response bias (e.g., participants might feel compelled to minimize—or, alternatively, exaggerate—their experience).[5]

Debriefings and post-study interviews would be respectful to participants. If results are collected and disseminated, they would also contribute to knowledge about risks and harm in SBES research. In turn, such knowledge could help OHRP formulate guidance for IRBs on such practices as exemption and use of expedited review and thereby facilitate research by reducing the variability in how IRBs handle these aspects of review. Toward this goal, we urge funding agencies and OHRP to encourage researchers to build participant assessments of risks and harm into their study designs.

ONGOING DATA SYSTEM

Recommendation 6.4: The Office for Human Research Protections should establish an ongoing system for collecting and publishing data that can help assess how effectively IRBs protect human research participants, how efficiently they review research, and how commensurate review is with risk.

There is astonishingly little hard information about the operation of the IRB system for human research participant protection in the United States today and how IRBs are interpreting provisions of the Common Rule. Research institutions are required to provide information about IRB membership to OHRP in order to obtain a federal-wide assurance (FWA) that authorizes the IRB to operate.[6] IRBs must also retain their meeting minutes and other records for a specified period, and they must report any instances that investigators report to them of harm to a research participant (adverse impact reports). However, IRBs are not required to submit regular reports of activities, such as meetings held, protocols reviewed, disposition of protocols, or other

---

[5] Participant debriefings are not meant to substitute for the reports of specific harms that investigators are required to make to IRBs upon occurrence—for example, if a survey respondent were to experience a panic attack in response to questioning, the investigator should report that harm immediately to the IRB. In turn, the IRB is required to report specific harms to OHRP.

[6] Previously, institutions had to provide extensive documentation of IRB procedures in order to obtain a multiple project assurance. The FWA procedure is designed to reduce the burden of assuring that an institution will comply with the Common Rule.

information that would provide a basis for monitoring their workloads and decisions.[7]

Furthermore, over the 35-year life of IRBs (and predecessor review committees), only a handful of major surveys have examined their characteristics, performance, and effects on human participants and research projects. Findings from these surveys are not often comparable, and most findings are not reported by type of research.

The most recent comprehensive survey of IRB operations was in 1995, and it has significant deficiencies, including a restricted study universe and the absence of such important information as comparisons of informed consent procedures and forms before and after IRB review; comparisons of IRB and investigator assessments of harm and risk; breakdowns of results by specific field of research; and any harm reported for participants and how it relates to type of IRB review, such as full board or expedited review. Moreover, and, in our view, inexcusably, complete documentation for the Bell survey is not available because of lack of funding. Other surveys and case studies reported in the literature are even more out of date and limited in focus (e.g., surveying particular types of IRBs, research, or investigator disciplines; see Appendix D).

The limited available evidence, when carefully analyzed, is sufficient in our view to support a conclusion that IRBs are indeed overloaded and, furthermore, that some of the overload is the result of unnecessary resources being spent on minimal-risk research. Yet substantially better information would be required to monitor and refine the operations of the IRB system in the future.

We see a need for an ongoing federally funded data collection program that will provide regularly updated useful information on IRB characteristics, operating procedures, and outcomes. (Such a program is in addition to the systematic research that we recommend researchers undertake about perceptions and experiences of risks and harm by research participants.) Our sister Committee on Assessing the System for Protecting Human Research Participants also recognized this need. It asserted, "a fact that has repeatedly confounded this committee's deliberations is the lack of data regarding the scope and scale of current protection activities" (Institute of Medicine, 2002:4). It called for DHHS to "harmonize safety monitoring guidance for re-

---

[7]Some agencies have captured basic information, such as dollar funding and number of participants, on research they support that involves human participants. For example, the U.S. Department of Energy has a database on 300 such projects conducted or sponsored by the department since 1994 (see http://www.eml.doe.gov.hsrd [4/10/03]). However, there is no systematic database on the thousands of projects using human participants that are sponsored by DHHS.

search organizations, including standard practices for defining and reporting adverse events," "issue a yearly report summarizing the results of research monitoring activities in the United States, including OHRP and FDA findings from inspections conducted the previous year," "commission studies to gather baseline data on the current national system of protections for research participants," and "assemble data on the incidence of research injuries and conduct economic analyses of their costs."

We believe an OHRP data system on IRBs should include several types of indicators, including basic descriptive characteristics (e.g., number of members, disciplines represented, number of initial exemptions, expedited reviews, full board reviews); measures that could be used to assess performance (e.g., average elapsed time to complete action on protocols by type of research and type of review, percentage of expeditable and exemptible research that is expedited and exempted); and outcome measures (e.g., reported harms and reasons for their occurrence).

This data system would have its greatest positive effects on human participant protection and researcher conduct if it were viewed as the monitoring device for a large production process—in this case, the process of reviewing protocols by IRBs. By design, OHRP seeks some variation, reflective of community standards variation, and it does not want to increase threats to human participants by overly loose review nor to impede the progress of science by overly harsh prohibitions of research. The data system might provide a tool to measure both the average behavior of IRBs and the extent to which some IRBs are departing from that typical behavior. When the averages suggest too loose or too harsh reviews are occurring, OHRP might seek to clarify guidelines. When the variation in behavior across IRBs becomes too large to be reflective only of differences in community standards, OHRP might seek to learn why the IRBs showing unusual behavior are doing so.

For example, imagine the simple display of the distribution of the proportion of protocols that receive expedited review that appears in Figure 6-1. The figure shows a hypothetical distribution of expedited review at two time points. At time 1, most IRBs are expediting between 50 and 80 percent of eligible protocols (the boxed area denotes this), but there are some IRBs that expedite as few as 20 percent and some that expedite as many as 90 percent. OHRP might view this variation as problematic, examine the outliers through informal conversations with IRB chairs, and clarify guidelines. At time 2, following the intervention by OHRP, the variation in IRB behavior is much lower, with most IRBs expediting between 60 and 70 percent of eligible protocols and the outliers expediting 30 and 80 percent. Such a change would

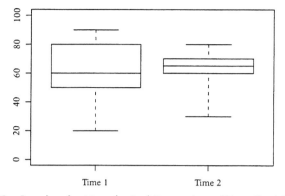

Figure 6-1 Boxplots for Hypothetical Proportion of Expedited Reviews Across IRBs

NOTES: These are hypothetical distributions for the proportion of protocols that would receive expedited review before and after a proposed change; see text for discussion.

signal to OHRP that the intervention led to more consistency across IRBs.

We recognize that it would be easy to interpret our recommendation as requiring the immediate imposition of burdensome paperwork requirements for every IRB. That is not our intention. At this time we propose that ORHP develop a relatively small-scale system which collects a limited set of data from a sample of IRBs, stratified by research volume at the institution. OHRP should consult with IRBs, researchers, and other relevant organizations to determine data priorities and the most feasible methods for obtaining the needed data items from sampled IRBs. As experience is gained with the system, it may be possible to expand its sample size and scope by taking full advantage of electronic communication technology.

IN-DEPTH STUDIES

Recommendation 6.5: Federal research funding agencies, including the National Science Foundation and the National Institutes of Health, should fund in-depth studies to better understand the operations and effects of the IRB system and to develop useful indicators of IRB performance.

It is not easy to devise performance measures for IRBs and other indicators that are feasible to collect, easy to define in a consistent manner among IRBs, and provide useful information. Moreover, for cost and feasibility reasons, an OHRP data system will necessarily be limited in the indicators that are regularly collected. For these reasons, we see a need for federal research funding agencies to fund in-depth studies on IRBs. Such research might use anthropological techniques, reviews of documents, and surveys (possibly including a longitudinal component) to examine various aspects of IRB functioning for different types of IRBs and kinds of research. The expectation is that a systematic research program would, in time, lead to better understanding that could inform the development of regularly collected performance measures and help guide policy for the operation of the IRB system.

As we noted in Chapter 4, the National Institutes of Health recently had a research program to study aspects of IRB operations, such as informed consent. This program should be continued and expanded. The National Science Foundation should similarly fund IRB-related in-depth research with a focus on SBES research.

— 7 —
System Issues

THE U.S. SYSTEM for human research participant protection involves many components—researchers, participants, research institutions, institutional review boards (IRBs), regulatory agencies, funding agencies, statistical agencies, professional associations, and others. The system is dynamic; it evolves as social, economic, and cultural changes affect various system components, and as they in turn respond—sometimes reactively, and sometimes with forethought and care about how to improve the operation of the system.

In the future as in the past the system should have two goals: first and foremost, the protection of the rights and welfare of research volunteers and, second, the facilitation of ethically responsible research that may result in useful knowledge about humans and human societies. The different actors in the system at times have different perspectives on how best to achieve these goals. Our report addresses some of these differing perspectives on the appropriate procedures for informed voluntary consent, methods and policies for protecting confidentiality, assessment of risk, harm, and benefit, and review procedures for minimal-risk research. Because of our charge and expertise, we have addressed these issues for research in the social, behavioral, and economic sciences (SBES); however, we believe that many of our recommendations are also relevant to biomedical research—and multidisciplinary research—that uses such methods as laboratory experiments, surveys, unstructured interviews, participant observation, and secondary analysis of existing data.

In this concluding chapter we consider system-level issues that we believe need continued attention. By system-level issues, we mean matters that involve the organization of components of the participant protection system and the relationships among actors in the system. The issues that we address fall into five broad categories: (1) guidance and support for IRBs; (2) qualifications and performance standards for IRBs and researchers; (3) communication among IRBs and researchers; (4) organization of and among IRBs; and (5) the development of national policy for human research participant protection. With limited time and resources, our discussion of these topics is limited. We highlight and endorse relevant recommendations of other groups and offer additional recommendations in a few areas that are particularly important for human participant protection in SBES research.

GUIDANCE AND SUPPORT FOR IRBS

Developing OHRP Guidance for IRBs

Throughout our report we stress the need for the Office for Human Research Protections (OHRP) to develop authoritative guidance for IRBs in several areas: appropriate ways to obtain and document informed consent for different types of SBES research (Chapter 4), good practices for confidentiality protection (Chapter 5), and effective review of minimal-risk research (Chapter 6). The process for developing such guidance will be easier in some areas than in others.

The most difficult area, we believe, relates to guidance on effective review of minimal-risk research—that is, guidance on applying the definitions of research and involvement of human subjects and when it is appropriate to exempt research or to conduct an expedited rather than full board review.[1] To be helpful and compelling for IRBs, such guidance needs to include concrete examples for a variety of research topics and methods.

It will not be easy to develop guidance for effective review of minimal-risk research in SBES fields for two reasons. The first impediment is the range of SBES research disciplines, topics, and methods, each of which presents somewhat different issues for human participant protection and the determination of minimal risk. A second impediment is that, unlike the case of some minimal-risk biomedical procedures (e.g., drawing blood below a specified amount), there are no evidence-based classifications of the risk level for specific SBES procedures (e.g., specific survey questions). Without such evidence, views may differ on, for example, whether and under what circumstances a survey question about alcohol use is minimal risk or more than minimal risk: When does such a question cause, at most, only temporary anxiety or embarrassment, and when may it cause longer lasting psychological trauma?

---

[1]Both the Institute of Medicine (IOM) committee and the National Bioethics Advisory Commission (NBAC) support the concept of review commensurate with risk: see Institute of Medicine (2002:Executive Summary) and National Bioethics Advisory Commission (2001:Rec.2.5). The IOM committee also recommended (Rec.3.3) that "the Office for Human Research Protections, with input from a broad spectrum of research disciplines and participant groups, should coordinate the development of guidance for risk classification."

Currently, IRBs are under pressure to be as risk-averse as possible. Without specific guidance, they are likely to assume the worst; however, in so doing, they may needlessly add to their workload and impede useful, ethically responsible research.

We suggest that OHRP, relevant professional associations, investigators, IRBs, and other interested groups regard the development of specific guidance as a long-term process that is carried out in a series of incremental steps.[2] Because the object is to develop guidance and not regulations, the process need not be as cumbersome or time-consuming as the regulatory process. One possible mode of operation is for OHRP to establish liaisons with relevant professional associations. Each association would be charged, in turn, to work with its members to develop specific examples to include in guidance about what is and is not research involving human subjects and what should be exempted or receive expedited review.[3] OHRP should issue such examples for public comment as soon as they are available, not waiting for all possible examples from all relevant associations.

It would also be very useful if SBES research funders were to sponsor research on risks posed by various SBES procedures, such as research on the circumstances in which participants believe that laboratory experiments or survey questions pose risks of temporary or longer-lasting harms of various types. Such research would be very helpful in enabling IRBs (and researchers) to make decisions about risk levels on the basis of science and not unsupported judgment (see also Recommendation 4.1 about the need for research on the effectiveness of alternative informed consent procedures and Recommendation 6.3 about the need for researchers to debrief participants about perceived and actual risks and harms).

Supporting IRBs

The human research participant protection system has developed as an add-on to the scientific research enterprise in the United States, without explicit recognition of the need to provide adequate financial support for IRB operations and adequate rewards for IRB service. Both the IOM committee and NBAC make strong recommendations

---

[2]Participant groups should also be involved in the development of guidance on minimal-risk and other pertinent issues, although participant involvement may be difficult to obtain because SBES research covers a wide range of populations—as distinct from a group of patients with a specific disease as is characteristic of much clinical research.

[3]Currently, SBES professional association guidelines on human research participant protection are addressed to researchers, not to IRBs.

for adequate funding for IRBs. The IOM report (Institute of Medicine, 2002:Rec. 2.3) states:

> Research sponsors and research organizations—public and private—should provide the necessary financial support to meet their joint obligation to ensure that Human Research Participant Protection Programs have adequate resources to provide robust protection to research participants.

NBAC is even more explicit. Its report recommends (National Bioethics Advisory Commission, 2001:Rec. 7.1) adequate funding for federal operations to oversee human research participant protection; inclusion of separate allocations in federal research programs for oversight activities; allowing research organizations to request funding for IRB and other protection activities; and additional funding from federal agencies, other research sponsors, and research organizations for IRB and other protection activities.

We agree with the intent of these recommendations, noting that there may be different mechanisms to achieve adequate funding for IRB and other protection activities (e.g., allowing direct charges to grants and contracts or allowing such charges to be included in indirect costs). In addition, we encourage research organizations to provide adequate financial and nonfinancial recognition for service on IRBs so that such service is viewed in a positive way and not as an unmitigated burden. In addition, universities and other research organizations should establish methods for evaluating the quality of IRB service and honor exemplary service by IRB members and chairs.

QUALIFICATIONS AND PERFORMANCE STANDARDS

Education and Training

Because of the work of many groups, including the National Science Foundation, the Office of Behavioral and Social Science Research in the National Institutes of Health (NIH), professional associations, individual IRBs, and individual researchers, there are now some materials available to help SBES researchers understand their responsibilities for human research participant protection and to learn how to navigate the maze of relevant regulations and procedures for project approval (see Appendix B; see also Oakes, 2002). There is much less guidance for IRBs or for researchers in other fields on how to handle different types of SBES research methods and populations studied. The available guidance for IRBs from OHRP and other sources is mostly general in nature, and training modules for researchers offered by NIH are oriented to clinical biomedical research. Education and training for both IRBs and researchers can be improved.

Looking to the future, we see three main concerns that are specific to SBES research. First, training materials are needed for IRBs that review SBES research. IRBs need in-depth discussions and analysis of the varieties of SBES research, issues that different types of research raise for human research participant protection, and effective ways to address those issues. The development of appropriate training materials for IRBs will require the joint efforts of OHRP, relevant SBES professional associations, and IRBs that are experienced in reviewing SBES research.[4]

Second, in the short term, training for SBES researchers of necessity must involve short courses at professional association meetings, on-line self-study modules, and the like. The training mechanisms should focus on specific issues of participant protection in different types of SBES research, such as appropriate consent processes for special populations, as much as possible. Such training should also reinforce the obligations of researchers to contribute to the participation protection system in every way possible—from submitting protocols for review that are ethically responsible and likely to contribute to useful knowledge to being willing to serve on IRBs.

Third, for the longer term, research institutions, particularly in their graduate programs, will need to develop ways to instill principles, practices, and responsibilities for ethical research as part of the basic professional education of SBES (and biomedical) researchers. The means to do so could include ethics modules as part of undergraduate courses, ethics courses as part of graduate training, and in-service courses for researchers to refresh understanding and address new developments in human research participant protection.

Many advisory bodies and commentators have highlighted the importance of education for IRBs and researchers in their recommendations (e.g., Advisory Committee on Human Radiation Experiments, 1996:Rec. 9; National Bioethics Advisory Commission, 2001:Rec. 3.1). For example, the Committee on Assessing the System for Protecting

---

[4]The Social and Behavioral Sciences Working Group, which is continuing to operate even though its parent National Human Research Protections Advisory Committee lost its charter in October 2002 (see Appendix B), is planning an activity in July 2003 on good practices for IRBs for the review of SBES research protocols. The activity will include a workshop followed by preparation of a document that is intended to help train IRB members (Felice Levine, American Educational Research Association, 2002, personal communication).

Human Research Participants of the Institute of Medicine (2002:Rec. 2.4) recommended:[5]

> Research organizations should ensure that investigators, Institutional Review Board members, and other individuals substantively involved in research with humans are adequately educated to perform their respective duties. The Office for Human Research Protections, with input from a variety of scholars in science and ethics, should coordinate the development and dissemination of core education elements and practices for human research ethics among those conducting and overseeing research.

Accreditation

One response to heightened concern about the adequacy of the IRB system for protecting human research participants has been a call for greater oversight of IRBs by the federal government (see, e.g., Office of Inspector General, 1998b; legislation introduced by Senator Kennedy). Another response has been for private groups to develop voluntary accreditation programs for IRBs.

The IOM committee devoted its entire first report to accreditation issues (Institute of Medicine, 2001). The committee's final report (Institute of Medicine, 2002) also contains several recommendations about accreditation, including Recommendation 6.4, which calls for continuation of efforts to develop voluntary accreditation programs:

> Voluntary accreditation should continue to be pilot tested as an approach to strengthening human research participant protections. The Department of Health and Human Services should arrange for a substantive review and evaluation of the accreditation process after five years, to be conducted under the purview of an independent entity.

The final report of the NBAC (2001) goes even further. It recommends not only that "Sponsors, institutions, and independent Institutional Review Boards should be accredited in order to conduct or review research involving human participants. ... " (Rec. 3.4), but also that "All investigators, Institutional Review Board members, and Institutional Review Board staff should be certified prior to conducting or

---

[5]The IOM report also recommends training for research participants so that they "understand their potential role in any study in which they enroll, the rationale underlying that study, and importantly, what is required of them to prevent unanticipated harm to themselves and to maintain the scientific integrity of the study" (Institute of Medicine, 2002:Rec. 4.2).

reviewing research involving human participants. . . . " (Rec. 3.3). The report asserts (p. xiv) that "Although accreditation and certification do not always guarantee the desired outcomes, these programs, which generally involve experts and peers developing a set of standards that represents a consensus of best practices, can be helpful in improving performance."

While not taking a position at this time on the benefits and costs of accreditation or certification programs, we support the IOM recommendation for continued testing and review of IRB accreditation programs. We offer two additional points. First, accreditation programs should involve researchers from the range of SBES (and biomedical) disciplines and take cognizance of appropriate review practices for different types of SBES research methods and populations studied. In this regard, it is encouraging to note that the Association for the Accreditation of Human Research Protection Programs added seats to its board for SBES researchers and has pilot-tested accreditation procedures at research institutions that conduct SBES research.

Second, accreditation programs need, so far as possible, to emphasize the spirit and not just the letter of the Common Rule regulations. For example, accreditation should focus on how IRBs review proposals to assure an appropriate process for obtaining informed consent and not just the documentation of such consent (see Chapter 4). In addition, accreditation programs should have standards for review commensurate with risk (see Chapter 6). Conducting reviews commensurate with risk helps IRBs allocate their limited resources to ensure protection of human participants while enabling responsible research to proceed. For example, a standard might be that IRBs should exempt a high proportion (specified range) of research that is eligible for exemption. Another standard might set different targets for the length of time to complete reviews depending on risk and the type of review conducted.

COMMUNICATION AMONG IRBS AND RESEARCHERS

IRB-Researcher Interaction

Recommendation 7.1: To improve IRB-researcher communication and facilitate the review process, IRBs should:

- clearly distinguish and justify changes to research protocols that are required for human participant protection from suggested changes that are advisory; and
- develop ways to work cooperatively with investigators, such as providing opportunities for face-to-face meet-

ings to discuss significant changes in research protocols that the IRB requires.

We believe that miscommunication between IRBs and researchers
may be one of the reasons that SBES researchers have often been frus-
trated with the IRB system, while IRBs, in turn, may sometimes believe that SBES researchers are not paying sufficient heed to human
participation protection requirements. Clear, open communication between IRBs and investigators is needed to facilitate the preparation
of research protocols that adequately describe participant protection
procedures and the timely review of research protocols by IRBs. To
the extent that researchers better understand the functions of and constraints on IRBs and IRBs better understand researchers' concerns for
maintaining the integrity of their research design and reaching closure
on a timely basis, the smoother the review process is likely to be.

Scientific Review

The Common Rule (45 *CFR* 46.111a) charges IRBs to determine for
each protocol that "risks to subjects are minimized: (i) By using procedures which are consistent with sound research design. ... " This
language is based on language first proposed by the Office for Protection from Research Risks (OPRR) in NIH in August 1979 (see Box A-6
in Appendix A). The OPRR proposal also included a provision for IRBs
to determine that "the research methods are appropriate to the objectives [of] the research and the field of study." This provision would have
considerably expanded the technical review function of IRBs, but it did
not survive the battle over the OPRR proposals (see Chapter 3).

Anecdotal evidence (including the experience of panel members)
suggests that IRBs sometimes require technical changes to SBES research designs that are not necessary for human participant protection
and that go beyond the expertise of the IRB members. For example,
an IRB may require a change in wording of a survey question without
fully understanding the purpose of the question or the research that
went into testing the proposed wording; the IRB may also overestimate
the risk that the question poses to human participants.

Quantitative evidence on this point is hard to find. According to the
1995 Bell survey (Bell, Whiton, and Connelly, 1998:Figure 40), IRB
chairs reported that they rarely criticized research designs; most criticisms instead were of consent forms. Similarly, investigators reported
that they were required to modify their proposed scientific design only
6 percent of the time and how they recruited participants only 11 percent of time. Most commonly (78 percent of the time), they were re-

quired to modify the consent form. In response to a subjective assessment, 56 percent of IRB chairs and 55 percent of IRB members agreed or strongly agreed that "this IRB's reviews improve the scientific quality of research done on human subjects," but only 37 percent of investigators shared this perception. None of the survey results is reported separately for SBES research.

From these data, it is possible that SBES researchers are overstating the propensity of IRBs to require inappropriate design changes that are not needed for human participant protection. Nevertheless, to minimize both the reality and the perception, we urge IRBs to clearly identify and justify changes that the IRB requires on behalf of human participant protection and to offer other suggestions on a purely advisory basis.[6]

Involvement of Investigators

The Common Rule does not specify how IRBs should communicate with investigators except to require an IRB to inform the investigator(s) "in writing of its decision to approve or disapprove the proposed research activity, or of modifications required to secure IRB approval of the research activity" (45 *CFR* 46.109d). The 1995 Bell survey found that 42 percent of low-volume IRBs routinely encouraged investigators to attend IRB meetings in person or to be reachable by telephone. In contrast, only 17 percent of high-volume IRBs followed this policy. Similarly, our review of IRB websites at 47 large universities found that only 15 percent invite attendance by investigators (another 9 percent have investigators sit outside the meeting to be available to answer questions).

We believe that IRBs should consider various ways to develop more open, less adversarial communications with researchers. Greater openness has the potential to facilitate understanding, resolve misunderstandings, improve the efficiency and timeliness of review, and build trust in the IRB system. Individual IRBs should consider the best means for improved communication, taking account of workload volume and other factors. Such means could include opening part of

---

[6]The IOM committee goes further, recommending that research organizations establish three related bodies for review of research protocols. One body would consider scientific issues; another body would consider financial conflict of interest; the two bodies would each make recommendations to a Research Ethics Review Board, which would have final approval authority but concentrate its own efforts on ethical issues (Institute of Medicine, 2002:Rec. 3.1, 3.2). We believe that such a structure requires careful consideration of its merits and costs before being considered for adoption. It is possible that such a structure could separate scientific review too much from ethical review or add more steps to the approval process.

IRB meetings to investigators (and, possibly, potential research participants) or scheduling face-to-face meetings for the IRB or IRB chair with the investigator to discuss significant changes to protocols.

Clear Guidance from IRBs to Researchers

To facilitate better communication of IRB expectations, so that researchers submit research protocols for review that fully address human participant protection issues, we believe that IRBs should provide clear guidance about what constitutes an acceptable protocol. A useful practice in this regard could be to post on websites outstanding examples of approved research protocols that meet high standards for participant protection in such areas as confidentiality protection and informed consent. Examples of protocols that were exempted or received expedited review, covering a variety of disciplines and methods used, could also be posted. Currently the IRB websites of major research universities often do not provide any more guidance on these matters than is contained in the Common Rule itself. For example, only 13 percent of these IRBs provide guidance for requesting exemption that does not simply repeat the Common Rule list of eligible categories of research, and only 11 percent provide guidance on confidentiality protection. However, over half of these IRBs provide guidance on informed consent, and 45 percent provide on-line training modules or guide books.[7]

Another potentially useful practice could be to publish, on an ongoing basis, the titles, names of principal investigators, and review classification (exempt, expedited, full board review) of projects approved by an IRB. Such information would allow researchers preparing protocols to identify work similar to theirs, contact the principal investigators, and otherwise capture the policies of their IRBs from past board decisions. Finally, IRBs should be clear when their standards for review are more stringent than the Common Rule.

Appeals Process

There is currently no provision in the Common Rule for investigators to appeal an adverse decision from an IRB. Although an investigator may argue back and forth with an IRB about changes to the research protocol, ultimately, the IRB's decision about what changes

---

[7]See Appendix D; we have not evaluated the relevance and usefulness of on-line training and guidance.

must be made to secure approval is final.[8] Backing up that authority is the provision in the Common Rule whereby an IRB may "suspend or terminate approval of research that is not being conducted in accordance with the IRB's requirements or that has been associated with unexpected serious harm to subjects" (45 CFR 46.113).

The purpose of lodging ultimate approval authority with an IRB was to insulate IRBs from pressures from institutional officials to approve research that might be important to the institution but that the IRB determined would not adequately protect participants. Finding ways to enable IRBs to resist pressures that could potentially compromise ethical judgments was and is a legitimate goal. However, some researchers believe that the lack of an appeals process gives IRBs too much power over the conduct of research. Further study is needed to find better ways to maintain IRB integrity and yet also allow researchers greater access to the decision process.

The panel discussed the desirability of recommending that a formal appeals process be added to the human participant protection process. Such a provision would provide a recourse to an investigator who believed that IRB-required changes would harm the scientific integrity of the proposed research but not provide added participant protection. Some panel members wanted to recommend an appeals procedure, but other panel members feared that such a procedure could add another layer of bureaucratic oversight to the IRB system. These members hope that implementation of Recommendation 7.1 to improve IRB-research interaction and other steps—such as clearer guidance from IRBs to researchers and more explicit guidance from OHRP to IRBs—could go a long way to alleviate the possible need for a formal appeals process.

ORGANIZATION OF AND AMONG IRBS

Organization and Staffing of IRBs

The Common Rule requires that IRBs have at least five members with varying backgrounds, including at least one scientist and one non-scientist, and at least one member not affiliated with the research institution, but it does not otherwise specify IRB composition, size, staffing, or other organizational features. The several thousand IRBs in the United States differ in their size, disciplines and expertise of members, level of staffing, and other resources for their operations. As one would

---

[8]The Common Rule states that "Research covered by this policy that has been approved by an IRB may be subject to further appropriate review and approval or disapproval by officials of the institution. However, these officials may not approve the research if it has not been approved by an IRB" (45 CFR 46.112).

expect, high-volume IRBs have more members, are supported by more hours of administrative staff, and make greater use of consultants, on average, compared with low-volume IRBs (see data from Bell, Whiton, and Connelly, 1998, in Chapter 2).

Some research institutions have more than one IRB for heavy workloads and to better align expertise of members with type of research reviewed. From our review of IRB websites of 47 major research institutions, 13 percent of these institutions have two IRBs, and 23 percent have three or more, including one institution that has half a dozen IRBs, each assigned to a specific research area, operating under the umbrella of an executive committee.

The Department of Veterans Affairs (VA) recently sponsored research to develop "optimal" staffing costs for operating human research participant protection programs at VA medical centers. From interviews with participant protection experts, data from the 1995 Bell survey (Bell, Whiton, and Connelly, 1998), and data from a VA system on number of studies involving human participants at VA centers, Wagner and Barnett (2000) developed cost models for a hypothetical medium-volume center (averaging about 370 total IRB reviews per year) and a hypothetical high-volume center (averaging about 1,380 total IRB reviews per year). The cost models considered staff requirements in light of workloads and assumed greater efficiencies in handling reviews by high-volume compared with low-volume IRBs. The models projected that a high-volume center should have four IRBs supported by an extensive staff; a medium-volume center should have two IRBs supported by somewhat fewer staff. The study also concluded that regional IRBs should be put in place to handle reviews for two or more low-volume centers to ensure that appropriate expertise and experience is available in the review process.

The VA study represents an interesting effort to cost out staffing levels for IRBs and to propose an optimal organization for IRBs at the VA's medical centers. However, the study noted that "an optimally staffed and funded IRB does not guarantee high-quality reviews," and it called for research on factors related to quality and how quality relates to cost (Wagner and Barnett, 2000:3). We agree that research on the relationship of quality, funding and staffing levels, and IRB organization is important. Without such research it is not obvious, for example, whether it is preferable to have separate IRBs for different research fields or to have a system in which there is some type of cross-disciplinary review of research protocols. Too narrow a focus for an IRB may lead it to be too uncritically accepting of the research protocols it reviews, but too wide a focus is likely to mean that few (if any) of the IRB members have appropriate subject expertise.

Similarly, research on the role that administrative staff should play in helping the IRB process (e.g., not only specific tasks, but also amount of discretionary judgment) would be useful. The need for these and similar research studies on IRB staffing and organization is a strong argument for the kinds of data that we recommend be gathered on IRB operations and on the risks and harms of different types of SBES (and biomedical) research and for the research that we recommend on developing appropriate performance or quality indicators for the IRB process (see Chapter 6).

Research Involving Multiple Sites

An increasing number of research protocols involve investigators and data collection at more than one site. This is true not only for biomedical research (e.g., clinical trials that may enroll participants at dozens of sites), but also for many kinds of SBES research. Examples include evaluation studies of the effects of the 1996 welfare reform act in several cities by consortia of researchers at different locations and ethnographic studies of school violence at multiple sites.

There are no easy answers to how best to protect human participants in multisite studies. The IRB system was established in part to permit local community input—the Common Rule requires that IRB membership be diverse in terms "of race, gender, and cultural backgrounds and sensitivity to such issues as community attitudes" (45 *CFR* 46.107a). The Common Rule also states that "each institution [in a multisite project] is responsible for safeguarding the rights and welfare of human subjects" (45 *CFR* 46.114).

Obtaining the concurrence of all involved IRBs can be time consuming and frustrating because of differences in IRB practices and standards for project approval. Even more important, the integrity of a multisite research design may be jeopardized if, because of differing IRB requirements, there is insufficient uniformity of procedures across all sites at which participants are recruited. The Common Rule permits organizations involved in cooperative research projects to "enter into a joint review arrangement, rely upon the review of another qualified IRB, or make similar arrangements for avoiding duplication of effort," although such arrangements require the concurrence of the relevant federal agency (45 *CFR* 46.114). The IOM report recommends streamlining the review of multisite clinical trials, but its recommendation does not really solve the problem of how to keep IRBs from second-guessing each other and declining to cede their review authority to another IRB.[9]

---

[9]The IOM recommendation states: "The review of multisite trials should be stream-

We do not have a ready solution to this problem. We believe it should be a high-priority issue for OHRP and the Secretary's Advisory Committee on Human Research Protections. Currently, there are efforts under way to establish regional IRBs that deal with multisite research in the region. For example, several universities and hospitals that frequently conduct multisite research projects may establish a single IRB to review those projects. Experience gained from these IRBs may, over time, be helpful in determining effective structures and procedures for review of multisite research.

DEVELOPING NATIONAL POLICY FOR HUMAN RESEARCH PARTICIPANT PROTECTION

Leadership for National Policy Development

Leadership in developing national policy and providing adequate oversight of human research participant protection is a fundamental obligation of the federal government given its role as a major research sponsor and its obligation under the Constitution to promote the general welfare of the population. Because of ongoing changes in society, cultural values, research techniques, and other factors, there is a continuing need for the government to review and modify, as appropriate, its policies, guidance, and oversight with respect to participant protection. In turn, the government needs to receive advice and recommendations on human protection issues from a continuing body that represents a range of relevant expertise and backgrounds and is in fact and in perception independent of political concerns or manipulation.

In this spirit, the IOM committee recommended (Institute of Medicine, 2002:Rec. 7.1):

> Congress should authorize and appropriate funding for a standing independent, multidisciplinary, nonpartisan expert Committee on Human Research Participant Protections whose membership would include the perspective of the research participant.

We support the IOM recommendation and suggest that ways to promote independence of such a committee is to give it a long-term charter and use staggered terms for members. Furthermore, it could be useful

---

lined, as allowed by current regulations. One primary scientific review committee and one primary Research Ethics Review Board should assume the lead review functions, with their determinations subject to acceptance by the local committees and boards at participating sites" (Institute of Medicine, 2002:Rec. 3.7).

to have members nominated by professional associations in biomedical and SBES disciplines, in addition to members appointed by the president and Congress.

NBAC did not speak to the issue of an appropriately constituted advisory committee, but it recommended that an independent agency be created to lead and coordinate federal oversight in this area (National Bioethics Advisory Commission, 2001:Rec. 2.2):

> To ensure the rights and welfare of all research participants, federal legislation should be enacted to create a single, independent federal office, the National Office for Human Research Oversight (NOHRO), to lead and coordinate the oversight system. This office should be responsible for policy development, regulatory reform, ... research review and monitoring, research ethics education, and enforcement.

We were not charged to consider and do not take a position on the proper form or location of a federal agency for human participant protection. However, we believe it likely that an independent agency, with responsibility for developing unified, comprehensive federal regulations and guidance (see NBAC, 2001:Rec. 2.3), and advised by a continuing, independent expert committee, would bring desirable qualities to federal activities in this area. Properly funded and organized, such an agency should be well positioned to provide leadership, involve experts from the full range of research disciplines, bring the views of research participants to bear, and protect federal policy in this area from partisan concerns.

Involvement of SBES Researchers in National Policy Setting

Recommendation 7.2: Any committee or commission that is established to provide advice to the federal government on human research participant protection policy should represent the full spectrum of disciplines that conduct research involving human participants. In particular, such a body should include members who represent the range of the social, behavioral, and economic sciences.

The IOM recommendation for an independent continuing advisory committee to the federal government on human research participant protection calls for the committee membership to be multidisciplinary. As the history of human participant protection policy in the United States indicates (see Chapter 3), SBES researchers have had significant input to the policy development process, but they have had to struggle to be heard. We believe strongly that, as a matter of course, the SBES research community should be involved along with the biomedical community in providing policy advice to the federal government. The benefits of such involvement would include not only increased support for and understanding of human participant protection policies among SBES researchers, but also useful cross-fertilization of ideas and knowledge between SBES and biomedical researchers about such topics as confidentiality protection and effective informed consent. Such cross-fertilization is increasingly important given the growing interdisciplinary nature of much research today. We note that the Secretary's Advisory Committee on Human Research Protections (for the U.S. Department of Health and Human Services) was chartered in October 2002, replacing the National Human Research Protections Advisory Committee. The members of the new committee were announced in January 2003. The committee includes two psychologists but does not represent other SBES disciplines nor does it include participant representation.

CONTINUING SYSTEM EVOLUTION

The U.S. human research participant protection system is multilayered, requires the cooperation of many different components (each with somewhat different perspectives), and is continually changing as various forces affect one or another part of the system. Such complexity and change can be unsettling for IRBs and others in the system who want clear guidance about how to meet their responsibilities for human participant protection. At present, IRBs are under pressure to take a legalistic approach by ignoring the flexibility in the Common Rule that permits informed consent procedures to be appropriately tailored for protection of specific populations or that permits review commensurate with risk. In the area of confidentiality protection, IRBs and researchers may not be fully aware of technology-driven changes that are increasing the risks of disclosure of information about participants.

We believe that the adoption of the recommendations in our report on effective informed consent, enhanced confidentiality protection, and review commensurate with risk will result in better guidance for IRBs and researchers on these topics and facilitate cooperative interaction and more consistent application of policies and procedures. We expect that implementation will be gradual as evidence is obtained and guidance is developed that gives IRBs assurance to move forward and that addresses other challenges to the system. For example, it will take time to develop a confidentiality protection certification program such as we recommend in Chapter 5, but, as it comes on line, the result should be greater access to research data that embodies appropriate, state-of-the-art protections to minimize the risk of harmful disclosure.

We also recognize that no report represents the last word on what should be done and that continued evaluation and modification of guidance (and, sometimes, regulations) will be needed in the future. To help guide the human research participant protection system as it moves forward and to enable IRBs and others to cope constructively with change, we see two critical needs. The first is for an ongoing data collection program on the operations of IRBs and the conduct of research that can inform policy and the public, as we recommend in Chapter 6. The second, as we recommend above, is for an ongoing independent advisory body for the federal government that can bring to bear key perspectives from researchers in all relevant fields, including SBES fields, as well as from participants. These two initiatives are important not only to help the development of appropriate policy and guidance, but also to increase public trust in the ability of the U.S. system to protect the many volunteers who make it possible to conduct research.

References

Advisory Committee on Human Radiation Experiments
1996 *Final Report of the Advisory Committee on Human Radiation Experiments.* New York: Oxford University Press.
American Association of University Professors
2001 Protecting human beings: Institutional review boards and social science research. *Academe* 87(3):55-67.
Annas, G.J., and M.A. Grodin, eds.
1992 *The Nazi Doctors and the Nuremberg Code: Human Rights in Human Experimentation.* New York: Oxford University Press.
Association of American Universities
2000 *Report on University Protections of Human Beings Who Are the Subjects of Research.* Report and recommendations from AAU's Task Force on Research Accountability. Washington, D.C.: Association of American Universities (June 28).
Barber, B.
1979 Some perspectives on the role of assessment of risk/benefit criteria in the determination of the appropriateness of research involving human subjects. In Appendix to *The Belmont Report: Ethical Principles and Guidelines for the Protection of Human Subjects of Research.* National Commission for the Protection of Human Subjects of Biomedical and Behavioral Research. Washington, D.C.: U.S. Government Printing Office.
Barber, B., J.J. Lally, J. Kakarushka, and D. Sullivan
1973 *Research on Human Subjects: Problems of Social Control in Medical Experimentation.* New York: Russell Sage Foundation.
Barnbaum, D.
2002 Making more sense of "minimal risk." *IRB: Ethics & Human Research* May-June:10-15.
Barnes, D.M., A.J. Davis, T. Moran, et al.
1998 Informed consent in a multi-cultural cancer patient population: Implications for nursing practice. *Nursing Ethics* 5:412-423.
Beauchamp, T.L., R.R. Faden, R.J. Wallace, Jr., and L. Walters, eds.
1982 *Ethical Issues in Social Science Research.* Baltimore, Md.: Johns Hopkins University Press.
Beecher, H.K.
1970 *Research and the Individual: Human Studies.* Boston: Little, Brown, and Company.
Bell, J., J. Whiton, and S. Connelly
1998 *Evaluation of NIH Implementation of Section 491 of the Public Health Service Act, Mandating a Program of Protection for Research Subjects.* Report prepared under a National Institutes of Health contract, N01-OD-2-2109. Washington, D.C.: U.S. Department of Health and Human Services.
Bernard, H.R., ed.
2000 *Handbook of Methods in Cultural Anthropology.* Walnut Creek, Calif.: Alta Mira Press.
2001 *Research Methods in Anthropology: Qualitative and Quantitative Approaches.* 3rd edition. Walnut Creek, Calif.: Alta Mira Press.

183

Blumberg, H.H., C. Fuller, and A.P. Hare
1974 Response rates in postal surveys. *Public Opinion Quarterly* 38:113-123.
Botkin, J.R.
2001 Protecting the privacy of family members in survey and pedigree research. *Journal of the American Medical Association* 285(2):207-211.
Brainard, J.
2001 The wrong rules for social science? *The Chronicle of Higher Education.* March 9.
Carrillo, J.E., A.R. Green, and J.R. Betancourt
1999 Cross-cultural primary care: A patient-based approach. *Annals of Internal Medicine* 130:829-834.
Centers for Disease Control and Prevention
1999 *Guidelines for Defining Public Health Research and Public Health Non-Research.* Atlanta: Centers for Disease Control and Prevention. Available: http://www.cdc.gov/od/ads/opspoll1.htm [4/3/03].
Chlebowski, R.T.
1984 How many protocols are deferred? One IRB's experience. *IRB* September/October:9-10.
Cleary, R.E.
1987 The impact of IRBs on political science research. *IRB* May/June:6-10.
Cooke, R.A., A.S. Tannenbaum, and B.H. Gray
1978 A survey of institutional review boards and research involving human subjects. Pp. 293-302 in *Report and Recommendations on Institutional Review Boards, Appendix.* National Commission for the Protection of Human Subjects of Biomedical and Behavioral Research. Washington, D.C.: U.S. Government Printing Office.
Czajka, J.L., and D. Kasprzyk
2002 *Limiting Disclosure in Public Use Microdata: Background for the Next Generation of Individual Tax Models.* Final report submitted to Internal Revenue Service. Washington, D.C.: Mathematica Policy Research, Inc. (December 6).
Dalenius, T.
1983 Informed consent or R.S.V.P. Chapter 4 in W.G. Madow, H. Nisselson, and I. Olkin, eds., *Incomplete Data in Sample Surveys, Vol. 3, Proceedings of the Symposium.* New York: Academic Press.
Davis, T.C., R.F. Holcombe, H.J. Berkel, S. Pramanik, and S.G. Divers
1998 Informed consent for clinical trials: A comparative study of standard versus simplified forms. *Journal of the National Cancer Institute* 90(9):668-674.
de Sola Pool, I.
1979 Prior restraint. *The New York Times* December 16:E19.
1980 The new censorship of social research. *Public Interest* 59:56-68.
Doyle, P., J.I. Lane, J.J.M. Theeuwes, and L.V. Zayatz, eds.
2001 *Confidentiality, Disclosure, and Data Access: Theory and Practical Applications for Statistical Agencies.* Amsterdam: Elsevier North-Holland.
Dunne, T.
2001 Issues in the establishment and management of secure research sites. Chapter 14 in P. Doyle, J.I. Lane, J.J.M. Theeuwes, and L.V. Zayatz, eds., *Confidentiality, Disclosure, and Data Access: Theory and Practical Applications for Statistical Agencies.* Amsterdam: Elsevier North-Holland.
Ellikson, P.L., and J.A. Hawes
1989 Active vs. passive methods for obtaining parental consent. *Evaluation Review* 13:45-55.
Faden, R., and T.L. Beauchamp
1986 *A History and Theory of Informed Consent.* New York: Oxford University Press.

Federal Committee on Statistical Methodology
    1994 *Statistical Policy Working Paper No. 22: Report on Statistical Disclosure Limitation Methodology.* Washington, D.C.: U.S. Office of Management and Budget.
Fisher, C.B., and W.W. Tryon, eds.
    1990 *Ethics in Applied Developmental Psychology: Emerging Issues in an Emerging Field. Advances in Applied Developmental Psychology.* Vol. 4. Norwood, N.J.: Ablex.
Gates, G.W.
    2000 Confidentiality. Pp. 80-83 in M.J. Anderson, ed.-in-chief, *Encyclopedia of the U.S. Census.* Washington, D.C.: CQPress.
Goldman, J., and M.D. Katz
    1982 Inconsistency and institutional review boards. *Journal of the American Medical Association* 248:197-202.
Goldstein, A.O., P. Frasier, P. Curtis, A. Reid, and N.E. Kreher
    1996 Consent form readability in university-sponsored research. *Journal of Family Practice* 42(6):606-611.
Gray, B.H.
    1977 *Human Subjects in Medical Experimentation: A Sociological Study of the Conduct and Regulation of Clinical Research.* New York: Wiley-Interscience.
    1982 Regulatory context of social and behavioral research. Pp. 329-355 in T.L. Beauchamp, R.R. Faden, R.J. Wallace, Jr., and L. Walters, eds., *Ethical Issues in Social Science Research.* Baltimore, Md.: Johns Hopkins University.
Gray, B.H., R.A. Cooke, and A.S. Tannenbaum
    1978 Research involving human subjects. *Science* 201(4361):1094-1101.
Grodin, M.A., and L.H. Glantz, eds.
    1994 *Children as Research Subjects: Science, Ethics, and Law.* Oxford.
Grundner, T.M.
    1983 DHEW human subjects protection: The new regulations revisited. *Health Matrix* 1:37-41.
Gunn, P., A. Fremont, M. Bottrell, and L. Shugarman
    2002 The HIPPA Privacy Rule: Impact on Research Access to Health Information. Santa Monica, Calif.: RAND.
Gunsalus, C.K.
    2001 An examination of issues presented by proposals to unify and expand federal oversight of human subject research. Pp. D-1–D-27 in *Ethical and Policy Issues in Research Involving Human Participants, Vol. II, Commissioned Papers and Staff Analysis.* Bethesda, Md.: National Bioethics Advisory Commission.
Hammersley, M., and P. Atkinson
    1995 *Ethnography: Principles in Practice.* 2nd edition. Routledge.
Hauck, M., and M. Cox
    1974 Locating a random sample by random digit dialing. *Public Opinion Quarterly* 38:253-256.
Holliman, W.B., G.A. Soileau, J.M. Hubbard, and J. Stevens
    1986 Consent requirements and anxiety in university undergraduate students. *Psychological Reports* 59:175-178.
Hotz, V.J., R. Goerge, J. Balzekas, and F. Margolin, eds.
    1998 *Administrative Data for Policy-Relevant Research: Assessment of Current Utility and Recommendations for Development.* Report of the Advisory Panel on Research Uses of Administrative Data. Evanston, Ill.: Northwestern University/University of Chicago Joint Center for Poverty Research.
Humphreys, L.
    1975 *Tearoom Trade: Impersonal Sex in High Places.* New York: Aldine de Gruyter.

Institute of Medicine
2001 *Preserving Public Trust: Accreditation and Human Research Participant Protection Programs.* Committee on Assessing the System for Protecting Human Research Participants, Board on Health Sciences Policy. Washington, D.C.: National Academies Press.
2002 *Responsible Research: A Systems Approach to Protecting Research Participants.* Committee on Assessing the System for Protecting Human Research Participants, D.D. Federman, K.E. Hanna, and L. Lyman Rodriguez, eds., Board on Health Sciences Policy. Washington, D.C.: National Academies Press.

Institute for Social Research
1999 Protection of sensitive data: principles and practices for research staff. *Center Survey—A Staff Newsletter of the Survey Research Center* [University of Michigan] 9(4, April):1,3.

Johnson, C.
1982 Risks in the publication of fieldwork. Pp. 71-92 in J. Lieber, ed., *The Ethics of Social Research: Fieldwork, Regulation, and Publication.* New York: Springer-Verlag.

Jones, J.H.
1981 *Bad Blood.* New York: Free Press.

Kaufert, J.M., and A. O'Neil
1990 Analysis of a dialogue on risks in childbirth: Clinicians, epidemiologists, and Inuit women. Pp. 32-54 in S. Lindenbaum and M. Lock, eds. *Knowledge, Power and Practice: The Anthropology of Medicine in Everyday Life.* Berkeley, Calif.: University of California Press.

Kaufert, J.M., and R.W. Putsch
1997 Communication through interpreters in healthcare: Ethical dilemmas arising from differences in class, culture, language, and power. *The Journal of Clinical Ethics* 8(1):71-87.

Keiger, D., and S. De Pasquale
2002 Trials & tribulation. *The Johns Hopkins Magazine* 54(1):28-41.

Kelman, H.C.
1968 *A Time to Speak: On Human Values and Social Research.* San Francisco: Jossey-Bass.

Koenig, B.A., and J. Gates-Williams
1995 Understanding cultural difference in caring for dying patients. *Western Journal of Medicine* 163:244-9.

Marshall, P.A.
1992a Anthropology and bioethics. *Medical Anthropology Quarterly* 6(1):49-73.
1992b Research ethics in applied anthropology. *IRB: A Review of Human Subjects Research* 14(6, November-December):1-5.
2001 Informed consent in international health research. Pp. 101-134 in R. Levine, S. Gorovitz, and J. Gallagher, eds., *Biomedical Research Ethics: Updating International Guidelines: A Consultation.* Geneva: Council for International Organization of Medical Sciences, World Health Organization.

Marshall, P.A., and B.A. Koenig
1996 Anthropology and bioethics: Perspectives on culture, medicine, and morality. Pp. 347-373 in C. Sargeant and T. Johnson, eds., *Medical Anthropology: Contemporary Theory and Method.* 2nd edition. Westport, Conn.: Praeger Publishing Co.

Marshall, P.A., B.A. Koenig, P. Grifhorst, and M. van Ewijk
1998 Ethical issues in immigrant health care and clinical research. Pp. 203-226 in S. Louc, ed., *Handbook of Immigrant Health.* New York: Plenum Press.

McCarthy, C.R.
  1984 Introduction: The IRB and social and behavioral research. Chapter 3 in J.E. Sieber, ed., *NIH Readings on the Protection of Human Subjects in Behavioral and Social Science Research.* Frederick, Md.: University Publications of America, Inc.
  1998 The institutional review board: Its origins, purpose, function, and future. Chapter 16 in D.N. Weisstub, ed., *Research on Human Subjects: Ethics, Law, and Social Policy.* Oxford: Pergamon Press.
Milgram, S.
  1974 *Obedience to Authority.* New York: Harper & Row.
Moberg, D.P., and D.L. Piper
  1990 Obtaining active parental consent via telephone in adolescent substance abuse prevention research. *Evaluation Review* 14:315-323.
Moreno, J.D.
  2001 Protectionism in research involving human subjects. Paper I in *Ethical and Policy Issues in Research Involving Human Participants. Vol. II, Commissioned Papers and Staff Analysis.* Bethesda, Md.: National Bioethics Advisory Commision.
National Bioethics Advisory Commission
  2001 *Ethical and Policy Issues in Research Involving Human Participants. Vol. 1, Report and Recommendations.* Bethesda, Md.: National Bioethics Advisory Commission.
National Commission for the Protection of Human Subjects of Biomedical and Behavioral Research
  1978 *Report and Recommendations on Institutional Review Boards.* Washington, D.C.: U.S. Government Printing Office.
  1979 *The Belmont Report: Ethical Principles and Guidelines for the Protection of Human Subjects of Research.* Washington, D.C.: U.S. Government Printing Office. Available: http://ohrp.osophs.dhhs.gov/humansubjects/guidance/belmont.htm [4/10/03].
National Research Council
  1979 *Privacy and Confidentiality as Factors in Survey Response.* Committee on National Statistics. Washington, D.C.: National Academies Press.
  1986 *Sharing Research Data.* Committee on National Statistics, S.E. Fienberg, M.E. Martin, and M.L. Straf, eds. Washington, D.C.: National Academies Press.
  1993 *Private Lives and Public Policies: Confidentiality and Accessibility of Government Statistics.* Panel on Confidentiality and Data Access, G.T. Duncan, T.B. Jabine, and V.A. de Wolf, eds. Committee on National Statistics and Social Science Research Council. Washington, D.C.: National Academies Press.
  1997 *Assessing Policies for Retirement Income: Needs for Data, Research, and Models.* Panel on Retirement Income Modeling, C.F. Citro and E.A. Hanushek, eds. Committee on National Statistics. Washington, D.C.: National Academies Press.
  2000a *Improving Access to and Confidentiality of Research Data—Report of a Workshop.* Committee on National Statistics. Washington, D.C.: National Academies Press.
  2000b *Principles and Practices for a Federal Statistical Agency.* 2nd Edition. Committee on National Statistics, M.E. Martin, M.L. Straf, and C.F. Citro, eds. Washington, D.C.: National Academies Press.
National Science Foundation
  2002 Frequently Asked Questions and Vignettes: Interpreting the Common Rule for the Protection of Human Subjects for Behavioral and Social Science Research. Policy Office. Washington, D.C. Available: http://www.nsf.gov/bfa/dga/policy/hsfaqs.htm [4/10/03].

Norwood, J.L.
    1995   *Organizing to Count: Change in the Federal Statistical System.* Washington,
            D.C.: The Urban Institute Press.
Oakes, J.M.
    2002   Risks and wrongs in social science research: An evaluator's guide to the IRB.
            Division of Epidemiology, University of Minnesota. Forthcoming in *Evalua-
            tion Review.*
Office for Protection from Research Risks
    1993   *IRB Guidebook.* National Institutes of Health. Prepared by R.L. Penslar, Poyn-
            ter Center for the Study of Ethics and American Institutions, Indiana Uni-
            versity. Washington, D.C.: U.S. Department of Health and Human Services.
            Available: http://ohrp.osophs.dhhs.gov/irb/irb_buidebook.htm [5/15/03].
Office of Inspector General
    1998a   *Final Report on Low-Volume Institutional Review Boards.* Memo Report No.
            OEI-01-97-00194. Washington, D.C.: U.S. Department of Health and Human
            Services.
    1998b   *Institutional Review Boards: A Time for Reform.* Publication No. OEI-01-97-
            00193. Washington, D.C.: U.S. Department of Health and Human Services.
            Available: http://www.dhhs.gov/progorg/oei/reports/a275.pdf [4/3/03].
    2000   *Protecting Human Research Subjects—Status of Recommendations.* Publica-
            tion No. 0E1-97-00197. Washington, D.C.: U.S. Department of Health and
            Human Services.
Patullo, E.L.
    1982   Modesty is the best policy: the federal role in social research. Pp. 373-390 in
            T. L. Beauchamp, R.R. Faden, R.J. Wallace, Jr., and L. Walters, eds., *Ethical
            Issues in Social Science Research.* Baltimore, Md.: Johns Hopkins University.
Putsch, R.W. III
    1985   Cross-cultural communication: The special case of interpreters in health care.
            JAMA: Journal of the American Medical Association 254(23):3344-3348.
Reamer, F.G.
    1979   Protecting research subjects and unintended consequences: The effect of guar-
            antees of confidentiality. *Public Opinion Quarterly* 43:497-506.
Reiss, A., Jr.
    1979   Selected issues in informed consent and confidentiality with special reference
            to behavioral, social science research inquiry. In Appendix to *The Belmont
            Report: Ethical Principles and Guidelines for the Protection of Human Subjects
            of Research.* National Commission for the Protection of Human Subjects of
            Biomedical and Behavioral Research. Washington, D.C.: U.S. Government
            Printing Office.
Salvo, J.J.
    2000   Census tracts. Pp. 66-68 in M.J. Anderson, ed.-in-chief, *Encyclopedia of the
            U.S. Census.* Washington, D.C.: CQPress.
Seastrom, M.M.
    2001   Licensing. Chapter 11 in P. Doyle, J.I. Lane, J.J.M. Theeuwes, and L.V. Za-
            yatz, eds., *Confidentiality, Disclosure, and Data Access: Theory and Practical
            Applications for Statistical Agencies.* Amsterdam: Elsevier North-Holland.
Seltzer, W., and M. Anderson
    2002   NCES and the Patriot Act: An Early Appraisal of Facts and Issues. Presented
            at the Joint Statistical Meetings, New York, N.Y. Fordham University and Uni-
            versity of Wisconsin-Milwaukee (August).
Shea, C.
    2000   Don't talk to the humans: The crackdown on social science research. *Lingua
            Franca* 10:26-34.

Sieber, J.E.
1992 *Planning Ethically Responsible Research—A Guide for Students and Internal Review Boards*. Applied Social Research Methods Series, Vol. 31. Newbury Park, Calif.: Sage Publications.
2001 Summary of Human Subjects Protection Issues Related to Large Sample Surveys. Prepared for the Bureau of Justice Statistics, U.S. Department of Justice. Department of Psychology, California State University at Hayward.
Sieber, J.E., and R.M. Baluyot
1992 A survey of IRB concerns about social and behavioral research. *IRB: A Review of Human Subjects Research* 14(2):9-10.
Sieber, J.E., S. Plattner, and P. Rubin
2002 How (not) to regulate social and behavioral research. *This Week's News and Reflections*. July. The Federation of Behavioral, Psychological, and Cognitive Sciences, Washington, D.C.
Singer, E.
1978a Informed consent: Consequences for response rate and response quality in social surveys. *American Sociological Review* 43:144-162.
1978b The effect of informed consent procedures on respondent reactions to surveys. *Journal of Consumer Research* 5:4957.
1984 Public reactions to some ethical issues of social research: Attitudes and behavior. *Journal of Consumer Research* 11:501-509.
1993 Informed consent and survey response: A summary of the empirical literature. *Journal of Official Statistics* 9(2):361-375.
2001 Public perceptions of confidentiality and attitudes toward data sharing by federal agencies. Chapter 14 in P. Doyle, J.I. Lane, J.J.M. Theeuwes, and L.V. Zayatz, eds., *Confidentiality, Disclosure, and Data Access: Theory and Practical Applications for Statistical Agencies*. Amsterdam: Elsevier North-Holland.
2003 Exploring the meaning of consent: Participation in research and beliefs about risks and benefits. Survey Research Center, University of Michigan. Forthcoming in *Journal of Official Statistics*.
Singer, E., and M.R. Frankel
1982 Informed consent in telephone interviews. *American Sociological Review* 47:116-126.
Smelser, N.J., and P.B. Baltes, eds.
2001 *International Encyclopedia of the Social and Behavioral Sciences*. Oxford, England: Elsevier Science.
Smith, M.B.
1979 Some perspectives on ethical/political issues in social science research. Pp. 11-22 in M.L. Wax and J. Cassell, eds., *Federal Regulations: Ethical Issues and Social Research*. American Association for the Advancement of Science. Boulder, Colo.: Westview Press.
Stanley, B., and J.E. Sieber, eds.
1991 *Social Research on Children and Adolescents: Ethical Issues*. Newbury Park, Calif.: Sage.
Steeh, C.
1981 Trends in nonresponse rates, 1952-1979. *Public Opinion Quarterly* 45:40-57.
Stout, D.
1999 U.S., citing safety, suspends human research aid at Duke. *The New York Times*, May 12.
Sugarman, J., D.C. McCrory, and R.C. Hubal
1998 Getting meaningful informed consent from older persons: astructured literature review of empirical research. *Journal of the American Gerontological Society* 46:517-524.

Sugarman, J., D.C. McCrory, D. Powell, A. Krasny, B. Adams, E. Ball, and C. Cassell
1999 Empirical research on informed consent: an annotated bibliography. *The Hastings Center Report* 29(1, January/February):S1-S42.
Sweeney, L.
2001 Information explosion. Chapter 3 in P. Doyle, J.I. Lane, J.J.M. Theeuwes, and L.V. Zayatz, eds., *Confidentiality, Disclosure, and Data Access: Theory and Practical Applications for Statistical Agencies*. Amsterdam: Elsevier North-Holland.
Taylor, K.M., A. Bejak, and R.H.S. Fraser
1998 Informed consent for clinical trials: is simpler better? *Journal of the National Cancer Institute* 90(9):644-645.
Trice, A.D.
1987 Informed consent: VII, biasing of sensitive self-report data by both consent and information. *Journal of Social Behavior and Personality* 2:369-374.
Tropp, R.A.
1979 What problems are raised when the current DHEW regulation on protection of human subjects is applied to social science research? Pp. 18-1–18-17 in Appendix to *The Belmont Report: Ethical Principles and Guidelines for the Protection of Human Subjects of Research*. National Commission for the Protection of Human Subjects of Biomedical and Behavioral Research. Washington, D.C.: U.S. Government Printing Office.
1982 A regulatory perspective on social science research. Pp. 391-415 in T.L. Beauchamp, R.R. Faden, R.J. Wallace, Jr., and L. Walters, eds., *Ethical Issues in Social Science Research*. Baltimore, Md.: Johns Hopkins University.
U.S. General Accounting Office
1996 *Scientific Research: Continued Vigilance Critical to Protecting Human Subjects*. GAO/HEHS-96-72. Washington, D.C.: U.S. General Accounting Office.
Wagner, T.H., and P.G. Barnett
2000 *Human Subjects Compliance Programs: Optimal Operating Costs in VA*. VA Health Economics Resource Center Technical Report #2. Washington, D.C.: U.S. Department of Veterans Affairs.
Wax, M.L.
1979 Fieldwork, ethics, and politics. Pp. 85-102 in M.L. Wax and J. Cassell, eds., *Federal Regulations: Ethical Issues and Social Research*. American Association for the Advancement of Science. Boulder, Colo.: Westview Press.
Zimbardo, P.
1971 The Stanford Prison Experiment. Department of Psychology, Stanford University, Palo Alto, Calif.

Appendices

— A —
Tracing Changes in Regulatory Language

As an aid to following the changes in federal regulations for human research participant protection from 1974 through 1998, this appendix excerpts text from regulations and proposed regulations for human research participant protection on the following topics:

Box A-1, applicability of regulations;

Box A-2, definition of research;

Box A-3, definition of human subject;

Box A-4, research eligible for exemption;

Box A-5, expedited review (SBES-related categories);

Box A-6, criteria for IRB review;

Box A-7, basic elements of informed consent;

Box A-8, additional elements of informed consent;

Box A-9, conditions for waiver of informed consent;

Box A-10, documentation of informed consent and waiver conditions; and

Box A-11, definition of minimal risk.

As applicable, language is excerpted from 45 *CFR* 46, May 30, 1974; proposed regulations amending basic policy of the U.S. Department of Health, Education, and Welfare (HEW) (predecessor to the U.S. Department of Health and Human Services [HHS]), August 14, 1979; 45 *CFR* 46, January 26, 1981; 45 *CFR* 46, June 18, 1991; suggested revisions to the IRB expedited review list, November 10, 1997; expedited review list, November 9, 1998. Italics are added to note key differences from preceding or succeeding text.

BOX A-1
Applicability of IRB Regulations

May 30, 1974: 45 CFR 46, Protection of Human Subjects

46.1(a) The regulations in this part are applicable to all Department of Health, Education, and Welfare grants and contracts supporting research, development, and related activities in which human subjects are involved.

Aug. 14, 1979: Proposed Regulations Amending Basic HEW Policy for Protection of Human Research Subjects

46.101(a) Except as provided in paragraph (c), this subpart applies to all research involving human subjects conducted or supported by the Department of Health, Education, and Welfare.

46.122 Except for the categories of research exempted under 46.101(c), prior and continuing review and approval by an Institutional Review Board is required for the conduct of all research involving human subjects not funded by the Department, if the research is conducted at or supported by any institution receiving funds from the Department for the conduct of research involving human subjects.

Jan. 26, 1981: 45 CFR 46, Subpart A—Basic HHS Policy for Protection of Human Research Subjects

46.101(a) Except as provided in paragraph (b) of this section, this subpart applies to all research involving human subjects conducted by the Department of Health and Human Services or funded in whole or in part by a Department grant, contract, cooperative agreement or fellowship.

[proposed 46.122 dropped]

June 18, 1991: 45 CFR 46, Subpart A—Federal Policy for the Protection of Human Subjects

46.101(a) Except as provided in paragraph (b) of this section, this policy applies to all research involving human subjects conducted, supported, or otherwise subject to regulation *by any Federal Department or Agency* which takes appropriate administrative action to make the policy applicable to all such research.

BOX A-2
Definition of Research

May 30, 1974: 45 CFR 46, Protection of Human Subjects

No definition provided.

Aug. 14, 1979: Proposed Regulations Amending Basic HEW Policy...

46.102(e) "Research" means a formal investigation designed to develop or contribute to generalizable knowledge. Activities which meet this definition constitute "research" for purposes of this part, whether or not they are supported or conducted under a program which is considered research for other purposes. For example, some "demonstration" and "service" programs may include research activities.

Jan. 26, 1981: 45 CFR 46, Subpart A—Basic HHS Policy...

46.102(e) "research" means a *systematic* investigation designed to develop or contribute to generalizable knowledge. Activities which meet this definition constitute "research" for purposes of these regulations, whether or not they are supported or funded under a program which is considered research for other purposes. For example, some "demonstration" and "service" programs may include research activities.

June 18, 1991: 45 CFR 46, Subpart A—Federal Policy for the Protection of Human Subjects

46.102(e) "Research" means a systematic investigation, *including research development, testing and evaluation,* designed to develop or contribute to generalizable knowledge. Activities which meet this definition constitute research for purposes of this policy, whether or not they are conducted or supported under a program which is considered research for other purposes. For example, some demonstration and service programs may include research activities.

BOX A-3
Definition of Human Subject

May 30, 1974: 45 CFR 46, Protection of Human Subjects

No definition provided.

Aug. 14, 1979: Proposed Regulations Amending Basic HEW Policy. . .

46.102(f) "Human subject" means an individual about whom an investigator (whether professional or student) conducting research obtains (1) data through intervention or interaction with the person, or (2) identifiable information.

Jan. 26, 1981: 45 CFR 46, Subpart A—Basic HHS Policy. . .

46.102(f) "human subject" means a *living* individual about whom an investigator (whether professional or student) conducting research obtains (1) data through intervention or interaction with the individual, or (2) identifiable *private* information. [definitions of intervention, interaction, and private information follow]

June 18, 1991: 45 CFR 46, Subpart A—Federal Policy for the Protection of Human Subjects

[Same as Jan. 26, 1981]

BOX A-4
Research Eligible for Exemption

May 30, 1974: 45 CFR 46, Protection of Human Subjects

No provision to exempt any covered research.

Aug. 14, 1979: Proposed Regulations Amending Basic HEW Policy. . .

46.101(c) These regulations do not apply to:

Alternative A

(1) Research designed to study on a large scale: (A) the effects of proposed social or economic change, or (B) methods or systems for the delivery of or payment for social or health services.

(2) Research conducted in established or commonly accepted educational settings, involving normal educational practices, such as (A) research on regular and special education instructional strategies, or (B) research on the effectiveness of or the comparison among instructional techniques, curriculum, or classroom management.

(3) Research involving solely the use of standard educational diagnostic, aptitude, or achievement tests, if information taken from these sources is recorded in such a manner that subjects cannot be reasonably identified, directly or through identifiers linked to the subjects.

(4) Research involving solely the use of survey instruments if: (A) results are recorded in such a manner that subjects cannot be reasonably identified, directly or through identifiers linked to the subjects, or (B) the research (although not exempted under clause (A)) does not deal with sensitive topics, such as sexual behavior, drug or alcohol use, illegal conduct, or family planning.

(5) Research involving solely the observation (including observation by participants) of public behavior, if observations are recorded in such a manner that subjects cannot be reasonably identified, directly or through identifiers linked to the subjects.

(6) Research involving solely the study of documents, records, or pathological or diagnostic specimens, if information taken from these sources is recorded in such a manner that subjects cannot be reasonably identified, directly or through identifiers linked to the subjects.

(7) Research involving solely a combination of any of the activities described above.

Alternative B

(1) [same as Alternative A]

(2) [same as Alternative A]

(3) [same as Alternative A]

(4) *Survey activities involving solely product or marketing research, journalistic research, historical research, studies of organizations, public opinion polls, or management evaluations, in which the potential for invasion of privacy is absent or minimal.*

BOX A-4 (continued)

(5) Research involving the study of documents, records, data sets or human materials, when the sources or materials do not contain identifiers or cannot reasonably be linked to individuals. [similar to A.6]

(6) [same as Alternative A.7]

Jan. 26, 1981: 45 CFR 46, Subpart A—Basic HHS Policy...

46.101(b) Research activities in which the only involvement of human subjects will be in one or more of the following categories are exempt from these regulations unless the research is covered by other subparts of this part:

(1) Research conducted in established or commonly accepted educational settings, involving normal educational practices, such as (i) research on regular and special education instructional strategies, or (ii) research on the effectiveness of or the comparison among instructional techniques, curricula, or classroom management methods.

(2) Research involving the use of educational tests (cognitive, diagnostic, aptitude, achievement), if information taken from these sources is recorded in such a manner that subjects cannot be identified, directly or through identifiers linked to the subjects.

(3) Research involving survey or interview procedures, except where all of the following conditions exist: (i) Responses are recorded in such a manner that the human subjects can be identified, directly or through identifiers linked to the subjects, (ii) the subject's responses, if they become known outside the research, could reasonably place the subject at risk of criminal or civil liability or be damaging to the subject's financial standing or employability, and (iii) the research deals with sensitive aspects of the subject's own behavior, such as illegal conduct, drug use, sexual behavior, or use of alcohol. All research involving survey or interview procedures is exempt, without exception, when the respondents are elected or appointed public officials or candidates for public office.

(4) Research involving the observation (including observation by participants) of public behavior, except where all of the following conditions exist: (i) Observations are recorded in such a manner that the human subjects can be identified, directly or through identifiers linked to the subjects, (ii) the observations recorded about the individual, if they became known outside the research, could reasonably place the subject at risk of criminal or civil liability or be damaging to the subject's financial standing or employability, and (iii) the research deals with sensitive aspects of the subject's own behavior such as illegal conduct, drug use, sexual behavior, or use of alcohol.

(5) Research involving the collection or study of existing data, documents, records, pathological specimens, or diagnostic specimens, if these sources are publicly available or if the information is recorded by the investigator in such a manner that subjects cannot be identified, directly or through identifiers linked to the subjects.

BOX A-4 (continued)

June 18, 1991: 45 CFR 46, Subpart A—Federal Policy for the Protection of Human Subjects

46.101(b) Unless otherwise required by Department or Agency heads, research activities in which the only involvement of human subjects will be in one or more of the following categories are exempt from this policy:

(1) [same as Jan. 26, 1981, regulations]

(2) Research involving the use of educational tests (cognitive, diagnostic, aptitude, achievement), *survey procedures, interview procedures or observation of public behavior,* unless: (i) information obtained is recorded in such a manner that human subjects can be identified, directly or through identifiers linked to the subjects; *and (ii) any disclosure of the human subjects' responses outside the research could reasonably place the subjects at risk of criminal or civil liability or can be damaging to the subjects' financial standing, employability, or reputation.*

(3) Research involving the use of educational tests (cognitive, diagnostic, aptitude, achievement), survey procedures, interview procedures, or observation of public behavior that is not exempt under paragraph (b)(2) of this section, if: (i) *the human subjects are elected or appointed public officials or candidates for public office; or (ii) Federal statute(s) require(s) without exception that the confidentiality of the personally identifiable information will be maintained throughout the research and thereafter.*

(4) [same as Jan. 26, 1981, regulations, item 5]

(5) *Research and demonstration projects which are conducted by or subject to the approval of Department or Agency heads, and which are designed to study, evaluate, or otherwise examine: (i) Public benefit or service programs; (ii) procedures for obtaining benefits or services under these programs; (iii) possible changes in or alternatives to those programs or procedures; or (iv) possible changes in methods or levels of payment for benefits or services under those programs.*

(6) *Taste and food quality evaluation and consumer acceptance studies, (i) if wholesome foods without additives are consumed or (ii) if a food is consumed that contains a food ingredient at or below the level and for a use found to be safe, or agricultural chemical or environmental contaminant at or below the level found to be safe, by the Food and Drug Administration or approved by the Environmental Protection Agency or the Food Safety and Inspection Service of the U.S. Department of Agriculture.*

BOX A-5
Expedited Review (SBES-Related Categories)

May 30, 1974: 45 CFR 46, Protection of Human Subjects

No provision for expedited review.

Aug. 14, 1979: Proposed Regulations Amending Basic HEW Policy. . .

46.111(a) The Secretary will publish in the Federal Register a list of categories of research, involving no more than minimal risk, that may be reviewed by the Institutional Review Board through an expedited review procedure. The Secretary will amend this list, as appropriate, through republication in the Federal Register.

The Department proposes to include the following procedures in the list to be promulgated under this section: [SBES related categories only]

(6) Voice recordings made for research purposes such as investigations of speech defects.

(8) Program evaluation activities that entail no deviation for subjects from the normal requirements of their involvement in the program being evaluated or benefits related to their participation in such program.

Note.—The Department would add the following procedures to the above list if Alternative B under 46.101(c) is adopted: [see Box A-4 above]

(9) Survey activities to which responses are recorded in such a manner that individuals cannot reasonably be identified or in which the records will not contain sensitive information about the individuals.

(10) Research activities involving the observation of human subjects carrying out their normal day-to-day activities, where observations are recorded in such a manner that individuals cannot reasonably be identified.

(11) Research involving the study of documents, records, data sets, or human materials where the sources contain identifiers, but the researcher will take information from them in such a way as to prevent future identification of any individual.

Jan. 26, 1981: 45 CFR 46, Subpart A—Basic HHS Policy. . .

46.110(a) The Secretary has established, and published in the Federal Register, a list of categories of research that may be reviewed by the IRB through an expedited review procedure. The list will be amended, as appropriate, through periodic republication in the Federal Register.

46.110(b) An IRB may review some or all of the research appearing on the list through an expedited review procedure, if the research involves no more than minimal risk. The IRB may also use the expedited review procedure to review minor changes in previously approved research during the period for which approval is authorized. . .

BOX A-5 (continued)

Jan. 26, 1981: Research Activities Which May Be Reviewed Through Expedited Review Procedures. . . [SBES-related categories only]

(6) Voice recordings made for research purposes such as investigations of speech defects.

(8) The study of existing data, documents, records, pathological specimens, or diagnostic specimens.

(9) Research on individual or group behavior or characteristics of individuals, such as studies of perception, cognition, game theory, or test development, where the investigator does not manipulate subjects' behavior and the research will not involve stress to subjects.

June 18, 1991: 45 CFR 46, Subpart A—Federal Policy for the Protection of Human Subjects

46.110(a) [essentially the same as Jan. 26, 1981]

46.110(b) [essentially the same as Jan. 26, 1981]

Nov. 10, 1997: Suggested Revisions to the IRB Expedited Review List [SBES-related categories only]

(4) Research involving existing identifiable data, documents, records, or biological specimens (including pathological or diagnostic specimens) where these materials, in their entirety, have been collected prior to the research, for a purpose other than the proposed research.

(5) Research involving solely (a) prospectively collected identifiable residual or discarded specimens, or (b) prospectively collected identifiable data, documents, or records, where (a) or (b) has been generated for nonresearch purposes.

(7) Collection of data from voice, video, or image recordings made for research purposes where identification of the subjects and/or their responses would not reasonably place them at risk of criminal or civil liability or be damaging to the subjects' financial standing, employability, or reputation.

(8) Research on individual or group characteristics or behavior (including but not limited to research involving perception, cognition, surveys, interviews, and focus groups) as follows:

(a) Involving adults, where (i) the research does not involve stress to subjects, and (ii) identification of the subjects and/or their responses would not reasonably place them at risk of criminal or civil liability or be damaging to the subjects' financial standing, employability, or reputation.

(b) Involving children, where (i) the research involves neither stress to subjects nor sensitive information about themselves, or their family; (ii) no alteration or waiver of regulatory requirements for parental permission has been proposed; and (iii) identification of the subjects and/or their responses would not reasonably place them or their family members at risk of criminal or civil liability or be damaging to the financial standing, employability, or reputation of themselves or their family members.

BOX A-5 (continued)

Nov. 9, 1998: Categories of Research That May Be Reviewed by the IRB Through an Expedited Review Procedure [SBES-related categories only]

Applicability

(A) Research activities that (1) present no more than minimal risk to human subjects, and (2) involve only procedures listed in one or more of the following categories, may be reviewed by the IRB through the expedited procedure authorized by 45 CFR 46.110 and 21 CFR 56.110. The activities listed should not be deemed to be of minimal risk simply because they are included on this list. . .

(B) The categories in this list apply regardless of the age of subjects, except as noted.

(C) The expedited review procedure may not be used where identification of the subjects and/or their responses would reasonably place them at risk of criminal or civil liability or be damaging to the subjects' financial standing, employability, insurability, reputation, or be stigmatizing, unless reasonable and appropriate protections will be implemented so that risks related to invasion of privacy and breach of confidentiality are no more than minimal.

(D) The expedited review procedure may not be used for classified research involving human subjects.

(E) IRBs are reminded that the standard requirements for informed consent (or its waiver, alteration, or exception) apply regardless of the type of review—expedited or convened—utilized by the IRB.

(F) Categories one (1) through seven (7) pertain to both initial and continuing IRB review.

Research Categories

(5) Research involving materials (data, documents, records, or specimens) that have been collected or will be collected solely for nonresearch purposes (such as medical treatment or diagnosis). (Note: Some research in this category may be exempt. . . This listing refers only to research that is not exempt.)

(6) Collection of data from voice, video, digital, or image recordings made for research purposes.

(7) Research on individual or group characteristics or behavior (including, but not limited to, research on perception, cognition, motivation, identity, language, communication, cultural beliefs or practices, and social behavior) or research employing survey, interview, oral history, focus group, program evaluation, human factors evaluation, or quality assurance methodologies. (Note: Some research in this category may be exempt. . . This listing refers only to research that is not exempt.)

BOX A-6
Criteria for IRB Review

May 30, 1974: 45 CFR 46, Protection of Human Subjects

46.2(b) This review [by a committee of the organization receiving DHEW research funds] shall determine whether these subjects will be placed at risk, and, if risk is involved, whether:

(1) The risks to the subject are so outweighed by the sum of the benefit to the subject and the importance of the knowledge to be gained as to warrant a decision to allow the subject to accept these risks;

(2) the rights and welfare of any such subjects will be adequately protected;

(3) legally effective informed consent will be obtained by adequate and appropriate methods in accordance with the provisions of this part; and

(4) the conduct of the activity will be reviewed at timely intervals.

Aug. 14, 1979: Proposed Regulations Amending Basic HEW Policy. . .

46.110(a). . . In order to give its approval, the Board must determine that all of the following requirements are satisfied:

(1) *The research methods are appropriate to the objectives [of] the research and the field of study.*

(2) *Selection of subjects is equitable, taking into account the purposes of the research.*

(3) *Risks to subjects are minimized by using the safest procedures consistent with sound research design and, whenever appropriate, by using procedures already being performed for diagnostic or treatment purposes.*

(4) *Risks to subjects are reasonable in relation to anticipated benefits to subjects and importance of the knowledge to be gained. In making this determination, the Board should consider only those risks and benefits that may result from the research (as distinguished from risks and benefits the subjects would be exposed to or receive even if not participating in the research). Also, the Board should not consider possible effects of applying knowledge gained in the research as among those research risks which fall within the purview of its responsibility.*

(5) Informed consent will be sought from each prospective subject or his or her legally authorized representative, in accordance with and to the extent required by 46.112.

(6) *Informed consent will be appropriately documented, in accordance with, and to the extent required by, 46.113.*

(7) *Where appropriate, the research plan makes adequate provision for monitoring the data collected to ensure the safety of subjects.*

(8) *There are adequate provisions to protect the privacy of subjects and to maintain the confidentiality of data.*

BOX A-6 (continued)

(9) Applicable regulations for the protection of fetuses, pregnant women, children, prisoners, and those institutionalized as mentally disabled are satisfied.

Jan. 26, 1981: 45 CFR 46, Subpart A—Basic HHS Policy. . .

46.111(a) In order to approve research covered by these regulations the IRB shall determine that all of the following requirements are satisfied:

(1) Risks to subjects are minimized: (i) By using procedures which are consistent with sound research design and which do not necessarily expose subjects to risk, and (ii) whenever appropriate, by using procedures already being performed on the subjects for diagnostic or treatment purposes.

(2) Risks to subjects are reasonable in relation to anticipated benefits, if any, to subjects, and the importance of the knowledge that may reasonably be expected to result. In evaluating risks and benefits, the IRB should consider only those risks and benefits that may result from the research (as distinguished from risks and benefits of therapies subjects would receive even if not participating in the research). The IRB should not consider possible long-range effects of applying knowledge gained in the research (for example, the possible effects of the research on public policy) as among those research risks that fall within the purview of its responsibility.

(3) Selection of subjects is equitable. In making this assessment, the IRB should take into account the purposes of the research and the setting in which the research will be conducted.

(4) Informed consent will be sought from each prospective subject or the subject's legally authorized representative, in accordance with, and to the extent required by 46.116.

(5) Informed consent will be appropriately documented, in accordance with, and to the extent required by 46.117.

(6) Where appropriate, the research plan makes adequate provision for monitoring the data collected to insure the safety of subjects.

(7) Where appropriate, there are adequate provisions to protect the privacy of subjects and to maintain the confidentiality of data.

46.111(b) Where some or all of the subjects are likely to be vulnerable to coercion or undue influence, such as persons with acute or severe physical or mental illness, or persons who are economically or educationally disadvantaged, appropriate additional safeguards have been included in the study to protect the rights and welfare of these subjects.

June 18, 1991: 45 CFR 46, Subpart A—Federal Policy for the Protection of Human Subjects

46.111(a) [essentially same as Jan. 26, 1981, except for adding the following to item (3):

(3) . . . will be conducted and should be particularly cognizant of the special problems of research involving vulnerable populations, such as children, prisoners, pregnant women, mentally disable[d] persons, or economically or educationally disadvantaged persons.

BOX A-6 (continued)

46.111(b) When some or all of the subjects are likely to be vulnerable to coercion or undue, influence, such as *children, prisoners, pregnant women, mentally disabled persons,* or economically or educationally disadvantaged persons, additional safeguards have been included in the study to protect the rights and welfare of these subjects.

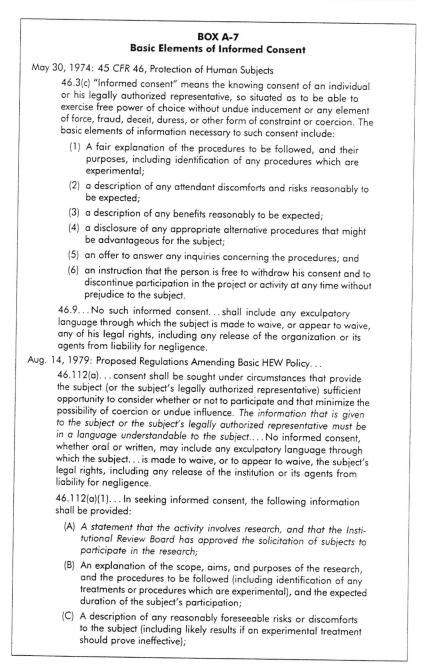

BOX A-7
Basic Elements of Informed Consent

May 30, 1974: 45 CFR 46, Protection of Human Subjects

46.3(c) "Informed consent" means the knowing consent of an individual or his legally authorized representative, so situated as to be able to exercise free power of choice without undue inducement or any element of force, fraud, deceit, duress, or other form of constraint or coercion. The basic elements of information necessary to such consent include:

(1) A fair explanation of the procedures to be followed, and their purposes, including identification of any procedures which are experimental;

(2) a description of any attendant discomforts and risks reasonably to be expected;

(3) a description of any benefits reasonably to be expected;

(4) a disclosure of any appropriate alternative procedures that might be advantageous for the subject;

(5) an offer to answer any inquiries concerning the procedures; and

(6) an instruction that the person is free to withdraw his consent and to discontinue participation in the project or activity at any time without prejudice to the subject.

46.9...No such informed consent...shall include any exculpatory language through which the subject is made to waive, or appear to waive, any of his legal rights, including any release of the organization or its agents from liability for negligence.

Aug. 14, 1979: Proposed Regulations Amending Basic HEW Policy...

46.112(a)...consent shall be sought under circumstances that provide the subject (or the subject's legally authorized representative) sufficient opportunity to consider whether or not to participate and that minimize the possibility of coercion or undue influence. *The information that is given to the subject or the subject's legally authorized representative must be in a language understandable to the subject.*...No informed consent, whether oral or written, may include any exculpatory language through which the subject...is made to waive, or to appear to waive, the subject's legal rights, including any release of the institution or its agents from liability for negligence.

46.112(a)(1)...In seeking informed consent, the following information shall be provided:

(A) *A statement that the activity involves research, and that the Institutional Review Board has approved the solicitation of subjects to participate in the research;*

(B) An explanation of the scope, aims, and purposes of the research, and the procedures to be followed (including identification of any treatments or procedures which are experimental), and the expected duration of the subject's participation;

(C) A description of any reasonably foreseeable risks or discomforts to the subject (including likely results if an experimental treatment should prove ineffective);

BOX A-7 (continued)

(D) A description of any benefits to the subject or to others which may reasonably be expected from the research;

(E) A disclosure of appropriate alternative procedures or courses of treatment, if any, that might be advantageous to the subject;

(F) *A statement that new information developed during the course of the research which may relate to the subject's willingness to continue participation will be provided to the subject;*

(G) *A statement describing the extent to which confidentiality of records identifying the subject will be maintained;*

(H) An offer to answer any questions the subject (or the subject's representative) may have about the research[,] the subject's rights, or related matters;

(I) *For research involving more than minimal risk, an explanation as to whether compensation and medical treatment are available if injury occurs and, if so, what they consist of or where further information may be obtained;*

(J) *Who should be contacted if harm occurs or there are questions or problems;* and

(K) A statement that participation is voluntary, refusal to participate will involve no penalty or loss of benefits to which the subject is otherwise entitled, and the subject may discontinue participation at any time without penalty or loss of benefits to which the subject is otherwise entitled.

Jan. 26, 1981: 45 CFR 46, Subpart A—Basic HHS Policy...

46.116 [Essentially the same as Aug. 14, 1979, 46.112(a)]

46.116 (a) Basic elements of informed consent. Except as provided in paragraph (c) of this section, in seeking informed consent the following information shall be provided to each subject:

(1) A statement that the study involves research, an explanation of the purposes of the research and the expected duration of the subject's participation, a description of the procedures to be followed, and identification of any procedures which are experimental.

(2) A description of any reasonably foreseeable risks or discomforts to the subject.

(3) [same as item D in Aug. 14, 1979, proposed regulations]

(4) [same as item E in Aug. 14, 1979, proposed regulations]

(5) A statement describing the extent, *if any,* to which confidentiality of records identifying the subject will be maintained.

(6) [essentially the same as item I in Aug. 14, 1979, proposed regulations]

(7) An explanation of whom to contact for answers to pertinent questions about the research and research subjects' rights, and whom to contact in the event of a research-related injury to the subject; and

BOX A-7 (continued)

(8) [same as item K in Aug. 14, 1979, proposed regulations]

June 18, 1991: 45 CFR 46, Subpart A—Federal Policy for the Protection of Human Subjects

46.116 [same as Jan. 26, 1981]

46.116 (a) Basic elements of informed consent. Except as provided in paragraph (c) or (d) of this section, in seeking informed consent the following information shall be provided to each subject:

[same as list in Jan. 26, 1991, regulations]

BOX A-8
Additional Elements of Informed Consent

May 30, 1974: 45 CFR 46, Protection of Human Subjects

No additional elements specified.

Aug. 14, 1979: Proposed Regulations Amending Basic HEW Policy...

46.112(a)(2): Additional elements. When appropriate, the Institutional Review Board shall require that some or all of the following elements of information also be provided:

(A) A statement that the particular treatment or procedure being tested may involve risks to the subject (or fetus, if the subject is pregnant or becomes pregnant) which are currently unforeseeable. This statement will often be appropriate in connection with tests of experimental drugs, or where the subjects are children, pregnant women, or women of childbearing age.

(B) Foreseeable circumstances under which the subject's participation may be terminated by the investigator without regard to the subject's consent.

(C) Any additional costs to the subject or others that may result from their participation in the research.

(D) Who is conducting the study, the approximate number of subjects involved, the institution responsible for the study, and who is funding it.

(E) The consequences of a subject's decision to withdraw from the research and procedures for orderly termination of participation by the subject.

Jan. 26, 1981: 45 CFR 46, Subpart A—Basic HHS Policy...

46.116(b): Additional elements of informed consent. When appropriate, one or more of the following elements of information shall also be provided to each subject:

(1) A statement that the particular treatment or procedure may involve risks to the subject (or to the *embryo* or fetus, if the subject is or may become pregnant), which are currently unforeseeable.

(2) [essentially the same as 46.112(a)(2) (B), Aug. 14, 1979, proposed regulations]

(3) Any additional costs to the subject that may result from their participation in the research.

(4) [same as 46.112(a)(2) (E), Aug. 14, 1979, proposed regulations]

(5) *A statement that significant new findings developed during the course of the research which may relate to the subject's willingness to continue to participate will be provided to the subject; and*

(6) *The approximate number of subjects involved in the study.*

BOX A-8 (continued)

June 18, 1991: 45 CFR 46, Subpart A—Federal Policy for the Protection of Human Subjects

46.116(b): Additional elements of informed consent. When appropriate, one or more of the following elements of information shall also be provided to each subject:

[essentially the same as in 46.116(b), Jan. 26, 1981, Basic HHS Policy]

BOX A-9
Conditions for Waiver of Informed Consent

May 30, 1974: 45 CFR 46, Protection of Human Subjects

No conditions specified.

Aug. 14, 1979: Proposed Regulations Amending Basic HEW Policy. . .

46.112(b): The Board may approve a consent procedure which does not include, or which alters, some or all of the elements of informed consent set forth in paragraph (a), provided the Board finds (and documents) the following:

(1) The withholding or altering will not materially affect the ability of the subject to assess the harm or discomfort of the research to the subject or others;

(2) Sufficient information will be disclosed to give the subject a fair opportunity to decide whether or not to participate;

(3) The research could not reasonably be carried out without the withholding or alteration;

(4) Information is not withheld or altered for the purpose of eliciting participation; and

(5) Whenever feasible the subject will be debriefed after his or her participation.

Jan. 26, 1981: 45 CFR 46, Subpart A—Basic HHS Policy. . .

46.116(c): An IRB may approve a consent procedure which does not include, or which alters, some or all of the elements of informed consent set forth above, *or waive the requirement to obtain informed consent* provided the IRB finds and documents that:

(1) *The research is to be conducted for the purpose of demonstrating or evaluating: (i) Federal, state, or local benefit or service programs which are not themselves research programs, (ii) procedures for obtaining benefits or services under these programs, or (iii) possible changes in or alternatives to these programs or procedures; and*

(2) The research could not practicably be carried out without the waiver or alteration.

46.116(d): An IRB may approve a consent procedure which does not include, or which alters, some or all of the elements of informed consent set forth above, *or waive the requirements to obtain informed consent* provided the IRB finds and documents that:

(1) *The research involves no more than minimal risk to the subjects;*

(2) The waiver or alteration will not adversely affect the rights and welfare of the subjects;

(3) The research could not practicably be carried out without the waiver or alteration; and

(4) Whenever appropriate, the subjects will be provided with additional pertinent information after participation.

BOX A-9 (continued)

46.116(e) The informed consent requirements in these regulations are not intended to preempt any applicable federal, state, or local laws which require additional information to be disclosed in order for informed consent to be legally effective.

June 18, 1991: 45 CFR 46, Subpart A—Federal Policy for the Protection of Human Subjects

46.116(c): An IRB may approve a consent procedure which does not include, or which alters, some or all of the elements of informed consent set forth above, or waive the requirement to obtain informed consent provided the IRB finds and documents that:

(1) the research or demonstration *project is to be conducted by or subject to the approval of state or local government officials* and is designed to study, evaluate, or otherwise examine: (i) public benefit service programs; (ii) procedures for obtaining benefits or services under these programs; (iii) possible changes in or alternatives to those programs or procedures; or (iv) possible changes in methods or levels of payment for benefits or services under those programs; and

(2) the research could not practicably be carried out without the waiver or alteration.

46.116(d) An IRB may approve a consent procedure which does not include, or which alters, some or all of the elements of informed consent set forth in this section, or waive the requirements to obtain informed consent provided the IRB finds and documents that:

[same four conditions as 46.116(d), Jan. 26, 1981, Basic HHS Policy]

46.116 (e) [same as 46.116(e), Jan. 26, 1981, Basic HHS Policy]

BOX A-10
Documentation of Informed Consent and Waiver Conditions

May 30, 1974: 45 CFR 46, Protection of Human Subjects

46.10. . . The documentation of consent will employ one of the following three forms:

(a) Provision of a written consent document embodying all of the basic elements of informed consent. This may be read to the subject or to his legally authorized representative, but in any event he or his legally authorized representative must be given adequate opportunity to read it. This document is to be signed by the subject or his legally authorized representative. Sample copies of the consent form as approved by the committee are to be retained in its records.

(b) Provision of a "short form" written consent document indicating that the basic elements of informed consent have been presented orally to the subject or his legally authorized representative. Written summaries of what is to be said to the patient are to be approved by the committee. The short form is to be signed by the subject or his legally authorized representative and by an auditor witness to the oral presentation and to the subject's signature. A copy of the approved summary, annotated to show any additions, is to be signed by the persons officially obtaining the consent and by the auditor witness. Sample copies of the consent form and of the summaries as approved by the committee are to be retained in its records.

(c) [to modify procedures (a) or (b) the review committee must establish]: (1) that the risk to any subject is minimal, (2) that use of either of the primary procedures for obtaining informed consent would surely invalidate objectives of considerable immediate importance, and (3) that any reasonable alternative means for attaining these objectives would be less advantageous to the subjects.

Aug. 14, 1979: Proposed Regulations Amending Basic HEW Policy. . .

46.113: Documentation of informed consent.

(a) Except as provided in paragraph (b), informed consent shall be documented in writing (and a copy provided to the subject or the subject's authorized representative) through either of the following methods:

[essentially the same as in 46.10 (a, b), May 30, 1974, Protection of Human Subjects]

(b) The Board may waive the requirement for the researcher to obtain documentation of consent for some or all subjects if it finds (and documents) either:

BOX A-10 (continued)

(1) That the only record linking the subject and the research would be the consent document, the only significant risk would be potential harm resulting from a breach of confidentiality, each subject will be asked whether he or she wants there to be documentation linking the subject with the research, and the subject's wishes will govern; or

(2) That the research presents no more than minimal risk of harm to subjects and involves no procedures for which written consent is normally required outside of the research context.

In many cases covered by this paragraph it may be appropriate for the Board to require the investigator to provide subjects with a written statement regarding the research, but not to request their signature, or to require that oral consent be witnessed.

Jan. 26, 1981: 45 CFR 46, Subpart A—Basic HHS Policy. . .

46.117 Documentation of informed consent.

(a) Except as provided in paragraph (c) of this section, informed consent shall be documented by the use of a written consent form approved by the IRB and signed by the subject or the subject's legally authorized representative. A copy shall be given to the person signing the form.

(b) Except as provided in paragraph (c) of this section, the consent form may be either of the following:

(1) A written consent document that embodies the elements of informed consent required by 46.116. This form may be read to the subject or the subject's legally authorized representative, but in any event, the investigator shall give either the subject or the representative adequate opportunity to read it before it is signed; or

(2) A "short form" written consent document stating that the elements of informed consent required by 46.116 have been presented orally to the subject or the subject's legally authorized representative. When this method is used, there shall be a witness to the oral presentation. Also, the IRB shall approve a written summary of what is to be said to the subject or the representative. Only the short form itself is to be signed by the subject or the representative. However, the witness shall sign both the short form and a copy of the summary, and the person actually obtaining consent shall sign a copy of the summary. A copy of the summary shall be given to the subject or the representative, in addition to a copy of the "short form."

(c) An IRB may waive the requirement for the investigator to obtain a signed consent form for some or all subjects if it finds either:

[essentially the same as 46.112(c), Aug. 14, 1979, Proposed Regulations Amending HEW Policy]

BOX A-10 (continued)

June 18, 1991: 45 CFR 46, Subpart A—Federal Policy for the Protection of Human Subjects

46.117 Documentation of informed consent.

[same as 46.117, Jan. 26, 1981, 45 CFR 46, Subpart A—Basic HHS Policy]

BOX A-11
Definition of Minimal Risk

May 30, 1974: 45 CFR 46, Protection of Human Subjects

46.3(b) "Subject at risk" means any individual who may be exposed to the possibility of injury, including physical, psychological, or social injury, as a consequence of participation as a subject in any research, development, or related activity which departs from the application of those established and accepted methods necessary to meet his needs, or which increases the ordinary risks of daily life, including the recognized risks inherent in a chosen occupation or field of service.

Aug. 14, 1979: Proposed Regulations Amending Basic HEW Policy. . .

46.102(g) "Minimal" risk is the probability and magnitude of harm that is normally encountered in the daily lives of *healthy* individuals, or in the routine medical, dental or psychological examination of healthy individuals.

Jan. 26, 1981: 45 CFR 46, Subpart A—Basic HHS Policy. . .

46.102 (g) "Minimal risk" means that the risks of harm anticipated in the proposed research are not greater, considering probability and magnitude, than those ordinarily encountered in daily life or during the performance of routine physical or psychological examinations or tests.

June 18, 1991: 45 CFR 46, Subpart A—Federal Policy for the Protection of Human Subjects

46.102 (i) "Minimal risk" means that the *probability and magnitude of harm or discomfort* anticipated in the research are not greater *in and of themselves* than those ordinarily encountered in daily life or during the performance of routine physical or psychological examinations or tests.

— B —
Selected Organizations and Resources for Human Research Participant Protection

Department of Health and Human Services (DHHS):

- **National Human Research Protections Advisory Committee (NHRPAC):** This Committee was established in June 2000 and disbanded in September 2002. Its role was to provide advice to the Office for Human Research Protections in DHHS. It was replaced in October 2002 by the Secretary's Advisory Committee on Human Research Protections. Information about NHRPAC and its activities can be found at http://ohrp.osophs.dhhs.gov/nhrpac/nhrpac.htm [4/10/03].

- **NHRPAC's Social and Behavioral Science Working Group (SBSWG):** One of SBSWG's major goals is to develop guidelines for the review of social and behavioral science research by institutional review boards (IRBs). This NHRPAC working group has addressed such issues as the review of public-use data files, risk and harm, and third parties. It is continuing its work independently of the new Secretary's Advisory Committee. It is planning an activity in July 2003 on best practices for IRBs for review of SBES research. The activity will include a workshop followed by preparation of a document to help train IRB members. Information is available at http://www.asanet.org/public/humanresearch [4/10/03].

- **National Institutes of Health (NIH) Office of Behavioral and Social Science Research (OBSSR):** The mission of OBSSR is to stimulate and integrate social and behavioral science research throughout NIH. OBSSR has produced a research agenda, which contains several topics germane to the protection of human research participants, entitled "Progress and Promise in Research on Social and Cultural Dimensions of Health: A Research Agenda,"

217

and maintains a website on IRB review of NIH-sponsored social and behavioral research. To learn more about OBSSR's products and activities, consult http://obssr.od.nih.gov [4/10/03].

- **Office for Human Research Protections (OHRP)**: OHRP is "[r]esponsible for overseeing human research subjects protections functions and related functions where research involves the use of human subjects."[1] This office was created in June 2000 when this responsibility was transferred from the NIH Office for Protection from Research Risks. Information is available at http://ohrp.osophs.dhhs.gov [4/10/03].

Executive Office of the President, Office of Science and Technology Policy (OSTP), Committee on Science, Human Subjects Research Subcommittee (HSRS): HSRS provides advice about interdepartmental issues in protection of human participants to OSTP's Committee on Science and to the departments and agencies that promulgate the "Common Rule." The chair of the Subcommittee is OHRP's director.

- **Non-Biomedical Working Group (NBMWG)**: The NBMWG was started in 2001 and is charged (1) with recommending or endorsing guidance or regulatory change to assist researchers, institutions, funding agencies, and research participants involved in nonbiomedical research; and (2) to work cooperatively with relevant advisory groups and other entities. This HSRS working group is addressing a variety of topics, including IRB review of public-use microdata and protection of the confidentiality of data. The NBMWG is co-chaired by Philip Rubin, National Science Foundation, prubin@nsf.gov, and Caroline Miner, Bureau of Prisons, cminer@bop.gov.

Federation of Behavioral, Psychological and Cognitive Sciences: The Federation is an association of scientific societies with interests in basic research on problems of behavior, psychology, language, education, knowledge systems and their psychological, behavioral, and physiological bases. On April 18 and 19, 2001, it convened a Forum on Research Management (FORM) issues in human research protection via the IRB. A summary of this FORM was posted on the Federation's website, http://federation.apa.org [4/10/03].

National Bioethics Advisory Committee (NBAC): NBAC was established in 1995 by President Clinton by Executive Order 12975. Its charter expired October 3, 2001. NBAC issued six reports. Its final

---

[1] DHHS, Office of the Secretary. *Federal Register*, 65(114), June 13, 2000, page 37136.

report, *Ethical and Policy Issues in Research Involving Human Participants*, issued in August 2001, is the most germane to the social, behavioral, and economic sciences. Georgetown University's National Reference Center for Bioethics Literature maintains the NBAC website, http://bioethics.georgetown.edu/nbac [4/10/03].

National Science Foundation's (NSF) Advisory Committee on Social, Behavioral, and Economic (SBE) Subcommittee for Human Subjects: This SBE Subcommittee held its initial meeting in June 2001. Its charge was to develop case studies and examples of the Common Rule that pertain to social, behavioral, and economic sciences research. These have been gathered in a workbook, and NSF posted this material on its website for use by researchers, IRB members, and relevant federal agencies (see National Science Foundation, 2002). For information on the NSF ad hoc subcommittee, contact Stuart Plattner, NSF, splattne@nsf.gov.

Public Responsibility in Medicine and Research (PRIM&R): Founded in 1974, this organization promotes the advancement of strong research programs and the consistent application of ethical precepts in both medicine and research. Through four national conferences per year and published reports from them, it has addressed a broad range of issues in biomedical and behavioral research, clinical practice, ethics, and the law, including the ethical and procedural issues surrounding IRBs; educating for the responsible conduct of research; the range of problems affecting AIDS research and treatment; reproductive and other technologies and their effects on patient care; health care ethics committees; scientific integrity and conflicts of interest; and the general range of questions surrounding academic/industrial relations. For information, see http://www.primr.org [4/10/03].

- Association for the Accreditation of Human Research Protection Programs (AAHRPP): This affiliate of PRIM&R was established in May 2001 as PRIM&R's national accrediting arm for protection programs. It is establishing a voluntary, peer-driven human research accreditation program, using a site visit model. For information, see http://www.aahrpp.org/principles.htm [4/10/03].

— C —
Agenda for Panel's First Meeting

AGENDA OF OPEN MEETING
June 20-21, 2001

11:15 a.m. Welcome and Introduction
Cora Marrett, *Chair*
Andrew White, *Director, Committee on National Statistics*
Christine Hartel, *Director, Board on Behavioral, Cognitive, and Sensory Sciences*

Institute of Medicine (IOM), Board on Health Sciences Policy, Committee on Assessing the System for the Protection of Human Research Participants, and its Sponsors
11:30 a.m. IOM's Committee on Assessing the System for the Protection of Human Research Participants
Roderick J.A. Little, *Member, IOM Committee; Chair, Department of Biostatistics, School of Public Health, University of Michigan*
11:45 a.m. U.S. Department of Health and Human Services, Office for Human Research Protections (OHRP)
Jeffrey M. Cohen, *Director, Division of Education and Development, OHRP*
12:00 p.m. U.S. Department of Health and Human Services, National Institutes of Health (NIH)
Belinda Seto, *Deputy Director, Office of Extramural Research, NIH*
12:15 p.m. Discussion
12:30 p.m. Lunch

Focus on Federal Statistical and Survey Organizations
1:30 p.m. U.S. Census Bureau
Gerald Gates, *Chief, Policy Office, Census Bureau*

1:45 p.m.   National Center for Education Statistics (NCES)
Jerry West, *Director, Early Childhood Studies Program, NCES*

2:00 p.m.   National Cancer Institute (NCI)
Gordon Willis, *Cognitive Psychologist, Applied Research Program, Division of Cancer Control and Population Sciences, NCI*

2:15 p.m.   Discussion

Focus on Professional Associations

2:30 p.m.   American Psychology Society (APS)
Barbara A. Spellman, *Associate Professor of Psychology, University of Virginia; Secretary, APS, and Co-Chair, APS Committee on Human Subject Protection*

2:45 p.m.   American Anthropological Association (AAA)
Mary Margaret Overbey, *Director of Government Relations, AAA, and David Guillet, Professor of Anthropology, Catholic University of America*

3:00 p.m.   American Political Science Association (APSA)
Robert J.P. Hauck, *Deputy Executive Director, APSA*

3:15 p.m.   American Psychological Association (APA)
Merry Bullock, *Associate Director for Science, APA, and Sangeeta Panicker, Research Ethics Officer, APA*

3:30 p.m.   Discussion

3:45 p.m.   Break

Focus on National Advisory Committees

4:15 p.m.   National Human Research Protections Advisory Committee (NHRPAC)

Overview of NHRPAC
Kate-Louise Gottfried, *Executive Director*
Overview of NHRPAC's Social and Behavioral Science Working Group
Felice Levine, *Cochair, Working Group (The second part of Dr. Levine's presentation will focus on the views of the American Sociological Association, where she serves as the Executive Director.)*

4:45 p.m.   National Bioethics Advisory Commission (NBAC)
Ellen Gadbois, *Senior Policy Analyst, NBAC*

5:00 p.m. National Science Foundation, Advisory Committee for Social, Behavioral, and Economic Sciences (SBE) Subcommittee for Human Subjects
        Norman Bradburn, *Assistant Director, SBE Directorate*

5:10 p.m. Open Discussion and Audience Comments

5:30 p.m. Reception

SPEAKERS

Norman Bradburn, National Science Foundation

Mary Bullock, Associate Director for Science, Science Directorate, American Psychological Association

Jeffrey Cohen, Director, Division of Education and Development, Office for Human Research Protections

Ellen Gadbois, Senior Policy Analyst, National Bioethics Advisory Commission

Gerald W. Gates, Chief, Policy Office, U.S. Census Bureau

Kate-Louise Gottfried, Executive Director, National Human Research Protections Advisory Committee

David Guillet, Department of Anthropology, The Catholic University of America

Christine Hartel, Director, Board on Behavioral, Cognitive, and Sensory Sciences, Division of Behavioral and Social Sciences and Education, National Research Council

Robert J.P. Hauck, Deputy Director, America Political Science Association

Felice Levine, Executive Director, American Sociological Association

Roderick J.A. Little, School of Public Health, University of Michigan

Mary Margaret Overbey, Director of Government Relations, American Anthropological Association

Sangeeta Panicker, Research Ethics Officer, Science Directorate, American Psychological Association

Belinda Seto, Deputy Director, Office of Extramural Research, National Institutes of Health

Barbara A. Spellman, Department of Psychology, University of Virginia

Jerry West, Director, Early Childhood Studies Program, National Center for Education Statistics

Gordon Willis, Cognitive Psychologist, Applied Research Program, Division of Cancer Control and Population Sciences, National Cancer Institute

Andrew White, Director, Committee on National Statistics, Division of Behavioral and Social Sciences and Education, National Research Council

INVITED GUESTS

Irma Arispe, National Center for Health Statistics
Nancy Bates, U.S. Census Bureau
Virginia S. Cain, National Institutes of Health
Lynda Carlson, National Science Foundation
Marcie Cynamon, National Center for Health Statistics
Anne Dierler, U.S. General Accounting Office
Nancy Donovan, U.S. General Accounting Office
John P. Fanning, U.S. Department of Health and Human Services
Brian Greenberg, Social Security Administration
Diane Hopkins, Westat, Rockville, MD
John Iceland, U.S. Census Bureau
Andrew Kessler, American Psychological Society
Dave Kleffman, National Institute of Justice
Jonathan Knight, American Association of University Professors
Alan Kraut, American Psychological Society
Dewey La Rochelle, Centers for Disease Control
Dev Mani, National Research Council
Tom McKenna, Westat, Rockville, MD
Caroline Miner, National Institute of Justice
Deborah Olster, National Science Foundation
Stuart Plattner, National Science Foundation
Michael Rand, U.S. Bureau of Justice Statistics
Holly Reed, National Research Council
Angela Sharpe, Consortium of Social Science Associations
Scott Spaulding, National Research Council
James Taggart, National Human Research Protections Advisory Committee; Johns Hopkins University
Ashley Trimmer, Social Science Research Council
Stanley Zimmerman, Westat, Rockville, MD

— D —
Selected Studies of IRB Operations: Summary Descriptions

Studies of institutional review boards (IRBs) are listed in chronological order; see References for full citations. Information is not available with which to evaluate the quality of individual studies.

Barber, B., J.J. Lally, J. Makarushka, D. Sullivan, 1973, "Research on Human Subjects: Problems of Social Control in Medical Experimentation"
This is the first known survey of IRBs, conducted in 1969; it included 300 biomedical IRBs, of which 70 percent had existed prior to the 1966 U.S. Public Health Service policy requirements. A single individual was interviewed at each institution. The survey found that relatively few IRBs required modifications of protocols: 34 percent had never modified or rejected a project.

Gray, Bradford H., Robert A. Cooke, Arnold S. Tannenbaum, 1978, "Research Involving Human Subjects"
This article in *Science* is a condensed version of the report of a study that the authors conducted for the National Commission for the Protection of Human Subjects of Biomedical and Behavioral Research under a contract to the University of Michigan (see Cooke, Gray, and Tannenbaum, 1978). The study covered a wide variety of issues and topics; highlights are excerpted below.
A probability sample of 61 institutions, stratified by type of institution and weighted by research volume, was drawn from a list of more than 420 institutions with IRBs approved by the U.S. Department of Health and Human Services; the survey covered research reviewed by IRBs at these institutions between July 1, 1974, and June 30, 1975. Interviews were conducted between December 1975 and July 1976 with

more than 2,000 research investigators whose proposals had been reviewed, over 800 IRB members, and almost 1,000 participants or third parties (e.g., parents) who consented on participants' behalf. The sample comprised IRBs at medical schools (including those that share an IRB with their university; 59 percent); separate university IRBs (18 percent); hospital IRBs (15 percent); and other. About 60 percent of studies reviewed were biomedical; about one-third were behavioral research; and the rest involved secondary analysis of data or tissue samples. IRBs varied widely in size, number of proposals reviewed per year, how often convened, and member-hours of IRB work per year. Fifty percent of IRB members were biomedical scientists; 21 percent were behavioral scientists.

Great diversity was evident in IRB procedures. For example, half had provisions for investigators to appeal IRB decisions, half did not; half assigned proposals to individual members for intensive review; half took formal votes; two-thirds accepted majority approval, one-fourth required unanimity; one-fourth said that investigators always attended meetings at which their projects were reviewed.

IRBs were active in modifying protocols: 50 percent of projects were modified or the IRB required more information; 44 percent were approved as is (6 percent had no data). These figures indicate greater activity by IRBs in modifying research than in the Barber et al. study (1973). In the Gray et al. study, 14 percent of the IRBs modified every study; 22 percent modified no more than one-third. The largest percentage of modifications involved consent forms and procedures.

One-fourth of investigators judged their projects to be without risk; one-fourth judged their projects to have no more than a "very low" probability of "minor" medical or psychological complications; the remainder judged their projects to involve a "low" probability of minor complications or a "very low" probability of "serious" complications.

IRBs often modified consent forms, but there were no significant differences in the average readability or completeness scores between the original and modified consent forms.

Those agreeing that "the review procedure has improved the quality of scientific research done at this institution—at least to some extent" were 78 percent of IRB biomedical members, 69 percent of biomedical researchers, 62 percent of IRB behavioral and social science members, and 55 percent of behavioral and social science researchers.

Those agreeing that "the review committee makes judgments that it is not qualified to make—at least to some extent" were 21 percent of IRB behavioral and social science members, 28 percent of IRB biomedical members, 43 percent of biomedical researchers, and 49 percent of behavioral and social science researchers.

High percentages (96-99 percent) of all four groups agreed that "the human subjects review procedure has protected the rights and welfare of human subjects—at least to some extent."

IRBs that made more requests for information or frequent modifications of protocols were likely to be viewed less favorably than less active IRBs.

Goldman, Jerry, and Martin D. Katz, 1982, "Inconsistency and Institutional Review Boards"

The investigators submitted three identical research protocols to 32 IRBs at major universities with medical colleges. Twenty-two IRBs participated by reviewing the three protocols and documenting their judgments. Each protocol contained serious ethical issues, had scientific design flaws, and provided an incomplete consent form. The participating IRBs varied in their judgments, and a substantial number approved the flawed designs. Although there was considerable consistency in objecting to the consent forms, IRBs did not identify specific deficiencies consistently.

Subsequently, the chair of one of the participating IRBs challenged Goldman and Katz, who responded in kind. This chair objected to the conclusion that IRBs needed more regulation so as to be more consistent. This chair also disagreed with the position of Goldman and Katz that it is part of the responsibility of IRBs to consider scientific design.

Chlebowski, Rowan T., 1984, "How Many Protocols Are Deferred? One IRB's Experience"

This study examined one IRB with jurisdiction over the clinical research program of a university-affiliated, major teaching hospital. The study reviewed the actions of this IRB on new protocols ($n = 168$) and continuing reviews ($n = 138$) over a 12-month period. Of new protocols, 27.9 percent were approved without comment, 64.4 percent were approved with comment, and 7.7 percent were deferred (none were disapproved). Reasons for deferral included: (1) study design was judged not likely to answer the research question, which made it difficult to determine the risk/benefit ratio; (2) concerns of lay members of IRB regarding adequacy of protection for participants; and (3) major problems with the consent form. Of 13 deferred proposals, 9 were resubmitted and approved, 4 were dropped. The IRB reached a unanimous decision in 97 percent of cases.

Cleary, Robert E., 1987, "The Impact of IRBs on Political Science Research"

This study surveyed 115 chairs of political science (PS) departments that offered Ph.D. degree programs and chairs of IRBs at these institutions. The response rate for political science chairs was 47 percent (54); the response rate for IRB chairs was 68 percent (78). For 30 institutions, responses were received from both PS and IRB chairs.

PS chairs reporting an IRB at their institution: 51 of 53.

PS chairs reporting experience with an IRB: 27 of 53.

Characterization of experience: 16 positive, 4 negative.

Social scientists on IRB: 25 yes, 4 no; PS members on IRB: 10 yes, 19 no.

IRB covered federally funded student research: 29 yes, 0 no; unfunded student research: 26 yes, 4 no.

IRB requires written advance informed consent: 26 yes, 4 no.

PS protocols cleared without change: 159; cleared after changes: 22; denied: 0.

Departments reporting significant problems with IRB: 3 yes, 26 no.

Chairs reporting IRB increased protection for human subjects: 19 yes, 6 no.

IRB chair responses largely agreed with department chair responses; 0 said that PS had problems with IRB, 54 said IRB had increased protection, 6 said it had not.

Problems identified included uncertainty and lack of information regarding informed consent and confidentiality; also, confusion at some institutions as to what research is covered, particularly unfunded student research. Rules also differed: one IRB required advance written consent for surveys; another specified verbal consent. Some institutions deleted the exemption for research with public officials or specified that such research had to be limited to official responsibilities of public officials.

Advisory Committee on Human Radiation Experiments, 1996, "Final Report"

Research Proposal Review Project:

This project sampled 125 research protocols involving human participants (84 involving ionizing radiation, and 41 not involving radiation) that were approved and funded by the Departments of Health and Human Services, Defense, Energy, and Veterans Affairs or by the National Aeronautics and Space Administration between fiscal years 1990 and 1993. Of the 84 radiation protocols, 31 were extramural (primarily from universities) and 53 were intramural. A committee member reviewed an additional 93 protocols with regard to informed consent issues.

The documents obtained about each of the 125 protocols were reviewed by two individuals, including at least one Advisory Committee member. The reviewers rated each protocol on ethical concerns, from 1 (no concerns) to 5 (serious concerns), and on level of risk (minimal, more than minimal). The reviewers also identified factors that resulted in poor ratings.

Overall ratings were distributed as follows: 1—34 percent; 2—34 percent; 3—18 percent, 4 or 5—14 percent. All studies receiving a 4 or 5 were also considered greater than minimal risk.

Factors contributing to poor ratings (3, 4, or 5) fell into three categories: (1) factors likely to affect how well potential participants understood the research and how they could benefit or be harmed (e.g., consent forms suggesting that participants might benefit from being treated by experimental drugs when such an outcome was highly unlikely); (2) factors likely to affect the voluntary nature of decisions about participation; and (3) approaches to the inclusion of people with limited or questionable decision-making capacity.

Subject Interview Study (SIS):

The SIS sample included almost 1,900 patients in oncology or cardiology clinics at medical institutions in fve areas of the country; all sampled patients received a brief interview, and 103 of them who were research participants received a longer interview.

The brief interview covered general attitudes toward medical research; understanding of such terms as clinical trial and medical experiment; beliefs about research participation; reasons for participating or not participating in research (when applicable); and demographic and other background information. The overall response rate was 95 percent.

Nearly 40 percent of patients had been research participants or invited to be participants. The attitudes of these patients were generally favorable to research; most felt free to decline or to leave the project.

Bell, James, John Whiton, and Sharon Connelly, 1998, "Evaluation of NIH Implementation of Section 491 of the Public Health Service Act, Mandating a Program of Protection for Research Subjects"

This is the most recent major study of IRBs. The study universe was defined as 491 IRBs that in 1995 operated with multiple project assurances under 45 *CFR* 46 and that had conducted more than 10 initial reviews of human participant research protocols in the previous year. Five groups received questionnaires: IRB chairs and institution officials at all 491 institutions; IRB administrators at 300 institutions; 4 investigators at each of the 300 institutions (1,200 investigators); 4 IRB members at each of 160 IRBs (640 members). Response rates were 80 percent or higher for IRB chairs (394), administrators (245), and institutional officials (400); rates were 68 percent for IRB members (435) and 53 percent for investigators (632).

Topics covered included:

- Person-time effort (total person-time of all IRB personnel, chair effort, member effort, administrator effort, institutional official effort, investigator effort on initial review);

- Effort per review (per initial review, per continuing review);

- Other information on effort (meeting time per review, duration of initial review, unimplemented protocols, multiple IRB reviews);

- Opinions about burden (overall efficiency, getting into inappropriate areas);

- General opinions and ratings relative to adequacy (rating of overall adequacy, effect of initial review on protocols, effect of IRB action versus other influences at the institution, relative effect of different IRB activities, effect on scientific quality, influence of workload on protection adequacy, bias/lack of expertise, investigators' ability to participate in review decisions, relative impact/burden of federal requirements);

- Concerns, modifications, and other review outcomes (approved as submitted, concerns raised in initial review, protocol modification, conditions on approval, actions on multicenter protocols);

- Other IRB actions (suspension or termination of approved research, overruling exemption determinations);

- Reports of potential problems (serious investigator noncompliance, within-jurisdiction harms, legal actions by subjects, subjects' complaints, informed consent process, problems with investigators);

- Suggested changes at the local level (enhancing IRB procedures and structure, education and training, additional resources);

- Suggested changes at the federal level (enhancing regulations and practices).

A few reported measures are distinguished by type of research (biomedical, SBES). A problem in evaluating the findings is that the sponsoring agency, the National Institutes of Health Office for Protection from Research Risks, never funded completion of the second volume of the study, which includes the questionnaires and other technical information. (Questionnaires were kindly provided to panel staff by James Bell Associates.)

Panel on Institutional Review Boards, Surveys, and Social Science Research, January 2003, Staff Review of University IRB Websites (unpublished)

Panel staff drew a sample of 48 of the 151 universities classified as "Doctoral/Research Universities-Extensive" in 2000 by the Carnegie Foundation for the Advancement of Teaching. These institutions typically offer a wide range of baccalaureate programs, and they are committed to graduate education through the doctorate. During 2000, they awarded 50 or more doctoral degrees per year across at least 15 disciplines. The final sample size was 47 institutions because one institution provided no information about IRBs on its website.

The staff examined IRB websites for the 47 institutions in December 2002–January 2003. These sites varied in the amount of information provided; also, some sites may have been more up to date than others. Based on the website materials only, the staff determined answers to the following questions:

1. How many IRBs does the university have?

1 IRB for the university	64%*
2 IRBs	13
3 or more (range is from 3 to 7)	23

* One university also has a College of Education Human Subjects Committee that can review unfunded projects involving no more than minimal risk.

2. Does the website list names of IRB members?

Yes	45%
No	55

3. Does the IRB require that undergraduate research be reviewed?

Yes	75%
Only if will be made public or is a thesis	6
Entire class, not individual projects	6
No	9
Could not determine	4

4. Does the IRB allow research to be exempted?

Yes	91%*
No	9

* At one university, department chairs make exemption decisions.

5. Is there guidance (beyond the Common Rule) for requesting exemption?

Yes	13%
No	79
Not applicable (does not exempt any research)	9

6. Does the IRB allow research to be expedited?

Yes	87%
No	13

7. Is there guidance (beyond the Common Rule) for requesting expedited review?

Yes	4%
No	83
Not applicable (does not expedite any research)	13

8. Is there guidance on informed consent, such as a template, that goes beyond the Common Rule?

Yes, guidance is appropriate for SBES research	64%
Yes, but guidance specifies inappropriate elements	9
Guidance simply repeats Common Rule	11
No guidance is provided	17

9. Is human participants protection training required for researchers?

Yes	57%
No	23
No, but there is a link to NIH on-line training	11
No, but there is a link to university on-line training	9

10. Are there on-line training modules and/or guide books?

Yes (developed by university), no SBES module	40%
Yes (developed by university), SBES module	4
Reference to NIH on-line training only	23
No	32

11. Is there guidance on confidentiality protection (beyond just stating a requirement)?

Yes	11%
No	89

12. What information is provided on how long the IRB review process will take?

Notice of frequency of IRB meetings	30%
Estimate of time to allow for review	43
No guidance provided	26
Minimum time for review for those providing time estimates ($n = 20$)	
At least 1 week	100%
At least 2 weeks	95
At least 3 weeks	85
At least 4 weeks	75
At least 6 weeks	30
Longer than 6 weeks	20

13. Is there guidance on minimal-risk research beyond the Common Rule definition?

Yes	0%
No	100

14. Is there any provision for investigators to meet with IRB face-to-face?

Yes, may attend meeting	15%
May wait outside meeting to answer questions	4
Meeting time and place listed, but no invitation	9
No	72

— E —
Confidentiality and Data Access Issues for Institutional Review Boards

George T. Duncan
Carnegie Mellon University

INTRODUCTION

A CCEPTED PRINCIPLES of information ethics (see National Research Council, 1993) require that promises of confidentiality be preserved and that the data collected in surveys and studies adequately serve their purposes. A compromise of the confidentiality pledge could harm the research organization, the subject, or the funding organization. A statistical disclosure occurs when the data dissemination allows data snoopers to gain information about subjects by which the snooper can isolate individual respondents and corresponding sensitive attribute values (Duncan and Lambert, 1989; Lambert, 1993). Policies and procedures are needed to reconcile the need for confidentiality and the demand for data (Dalenius, 1988).

Under a body of regulation known as the Federal Policy for the Protection of Human Subjects, the National Institutes of Health Office of Human Subjects Research (OHSR) mandates that institutional review boards (IRBs) determine that research protocols assure the privacy and confidentiality of subjects. Specifically, it requires IRBs to ascertain whether (a) personally identifiable research data will be protected to the extent possible from access or use and (b) any special privacy and confidentiality issues are properly addressed, e.g., use of genetic information. This standard directs an IRB's attention, but without elaboration and clarification it does not provide IRBs with operational criteria for evaluation of research protocols. Nor does it provide guidance to researchers in how to establish research protocols that can merit IRB approval. The Office for Human Research Protection (OHRP) is responsible for interpreting and overseeing implementation of the reg-

235

ulations regarding the Protection of Human Subjects (45 *CFR* 46) promulgated by the Department of Health and Human Services (DHHS). OHRP is responsible for providing guidance to researchers and IRBs on ethical issues in biomedical and behavioral research.

As IRBs respond to their directive to ethically oversee the burgeoning research on human subjects, they require systematic ways of examining protocols for compliance with best practice for confidentiality and data access. Clearly, the task of an IRB is lightened if researchers are fully aware of such practices and how they can be implemented.

This paper identifies key confidentiality and data access issues that IRB members must consider when reviewing protocols. It provides both a conceptual framework for such reviews and a discussion of a variety of administrative procedures and technical methods that can be used by researchers to simultaneously assure confidentiality protection and appropriate access to data.

CRITICAL ISSUES

Reason for Concern

Most generally, an ethical perspective requires researchers to maximize the benefits of their research while minimizing the risk and harm to their subjects. This beneficence notion is often interpreted that, first, "one ought not to inflict harm" and, second, that "one ought to do or promote good." In the context of assuring data quality from research studies, this means first assuring an adequate degree of confidentiality protection and then maximizing the value of the data generated by the research. Confidentiality is afforded for reasons of ethical treatment of research subjects, pragmatic grounds of assuring subject cooperation, and, in some cases, legal requirements.

Aspects of Concern

Data have a serious risk of disclosure when (a) disclosure would have negative consequences, (b) a data snooper is motivated—both psychologically and pragmatically—to seek disclosure (Elliot, 2001), and (c) the data are vulnerable to disclosure attack. Based on its confidentiality pledges, researchers must protect certain sensitive objects from a data snooper. Sensitive objects can be any of a variety of variables associated with a subject entity (person, household, enterprise, etc.). Examples include the values of numerical variables, such as household income, an X-ray of a patient's lung, and a subject's report of their sexual history. Data with particular characteristics pose substantial risk of disclosure and suggest vulnerability:

- geographical detail—census block (Elliot, Skinner, and Dale, 1998; Greenberg and Zayatz, 1992);

- longitudinal or panel structure—criminal histories (Abowd and Woodcock, 2001);

- outliers, likely unique in the population—such as a 16-year-old widow (Dalenius, 1986; Greenberg, 1990);

- attributes with high level of detail—income to the nearest dollar (Elliot, 2001);

- many attribute variables—such as medical record (Sweeney, 2001);

- population data, as in a census, rather than a survey with small sampling fraction (Elliot, 2001);

- databases that are publicly available, identified, and share individual respondents and attribute variables (*key variables*—Elliot and Dale, 1999) with the subject data—marketing and credit databases.

Data with geographical detail, such as census tract data, may be easily linked to known characteristics of respondents. Concern for this suggests placing minimum population levels for geographical identifiers. For particular geographical regions, this can mean specifying the minimum size of a region that can be reported. Longitudinal data, which tracks entities over time, also poses substantial disclosure risk. Many individuals had coronary bypass surgery in the Chicago area in 1998 and many had bypass surgery in Phoenix in 1999, but few did both. Outliers, say on variables like weight, height, or cholesterol level can lead to identifiable respondents. Data with many attribute variables allow easier linkage with known attributes of identified entities, and entities, which are unique in the sample, are more likely to be unique in the population. Population data pose more disclosure risk than data from a survey having a small sampling fraction. Finally, special concern must be shown when other databases are available to the data snooper and these databases are both identified and share with the subject data both individual respondents and certain attribute variables. Record linkage may then be possible between the subject data and the external database. The shared attribute variables provide the key.

Disclosure

The legitimate objects of inquiry for research involving human subjects are statistical aggregates over the records of individuals, for example, the median number of serious infections sustained by patients receiving a drug for treatment of arthritis. The investigators seek to provide the research community with data that will allow accurate inference about such population characteristics. At the same time, to respect confidentiality, the investigators must thwart the data snooper who might seek to use the disseminated data to draw accurate inferences about, say, the infection history of a particular patient. Such a capability by a data snooper would constitute a statistical disclosure.

There are two major types of disclosure—identity disclosure and attribute disclosure. *Identity disclosure* occurs with the association of a respondent's identity and a disseminated data record (Paass, 1988; Spruill, 1983; Strudler et al., 1986). *Attribute disclosure* occurs with the association of either an attribute value in the disseminated data or an estimated attribute value based on the disseminated data with the respondent (Duncan and Lambert, 1989; Lambert, 1993). In the case of identity disclosure, the association is assumed exact. In the case of attribute disclosure, the association can be approximate. Many investigators emphasize limiting the risk of identity disclosure, perhaps because of its substantial equivalence to the inadvertent release of an identified record. An attribute disclosure, even though it invades the privacy of a respondent, may not be so easily traceable to the actions of an agency. An IRB in its oversight capacity should be concerned that investigators limit the risk of both attribute and identity disclosures.

Risk of Disclosure

Measures of disclosure risk are required (Elliot, 2001). In the context of identity disclosure, disclosure risk can arise because a data snooper may be able to use the disseminated data product to reidentify some deidentified records. Spruill (1983) proposed a measure of disclosure risk for microdata: (1) for each "test" record in the masked file, compute the Euclidean distance between the test record and each record in the source file; (2) determine the percentage of test records that are closer to their parent source record than to any other source record. She defines the risk of disclosure to be the percentage of test records that match the correct parent record multiplied by the sampling fraction (fraction of source records released).

More generally, and consistent with Duncan and Lambert (1986, 1989), an agency will have succeeded in protecting the confidentiality of a released data product if the data snooper remains sufficiently uncertain about a protected target value after data release. From this perspective, a measure of disclosure risk is built on measures of uncertainty. Furthermore, an agency can model the decision making of the data snooper as a basis for using disclosure limitation to deter inferences about a target. Data snoopers are deterred from publicly making inferences about a target when their uncertainty is sufficiently high. Mathematically, uncertainty functions provide a workable framework for this analysis. Examples include Shannon entropy, which has found use in categorizing continuous microdata and coarsening of categorical data (Domingo-Ferrer and Torra, 2001; Willenborg and de Waal, 1996:138).

Generally, a data snooper has a priori knowledge about a target, often in the form of a database with identified records (Adam and Wortmann, 1989). Certain variables may be in common with the subject database. These variables are called *key* or *identifying* (De Waal and Willenborg, 1996; Elliot, 2001). When a single record matches on the key variables, the data snooper has a candidate record for identification. That candidacy is promoted to an actual identification if the data snooper is convinced that the individual is in the target database. This would be the case either if the data snooper has auxiliary information to that effect or if the data snooper is convinced that the individual is unique in the population. The data snooper may find from certain key variables that a sample record is unique. The question then is whether the individual is also unique on these key variables in the population. Bethlehem, Keller, and Pannekoek (1990) have examined detection of records agreeing on simple combinations of keys based on discrete variables in the files. Record linkage methodologies have been examined by Domingo-Ferrer and Torra (2001), Fuller (1993), and Winkler (1998).

Deidentification

Deidentification of data is the process of removing apparent identifiers (name, e-mail address, social security number, phone number, address, etc.) from a data record. Deidentification does not necessarily make a record anonymous, as it may well be possible to reidentify the record using external information. In a letter to DHHS, the American Medical Informatics Association (2000) noted:

> However, in discussions with a broad range of healthcare stakeholders, we have found the concept of "deidentified information" can be misleading, for it implies that if the

19 data elements are removed, the problem of reidentification has been solved. The information security literature
suggests otherwise. Additionally, with the continuing and
dramatic increase in computer power that is ubiquitously
available, personal health data items that currently would
be considered 'anonymous' may lend themselves to increasingly easy reidentification in the future. For these reasons,
we believe the regulations would be better served by adopting the conventions of personal health data as being of "High
Reidentification Potential" (e.g., the 19 data elements listed
in the current draft), and "Low Reidentification Potential."
Over time, some elements currently considered of low potential may migrate to the high potential classification. More
importantly, this terminology conveys the reality that virtually all personal health data has some confidentiality risk
associated with it, and helps to overcome the mistaken im-
pression that the confidentiality problem is solved by removing the 19 specified elements.

Most health care information, such as hospital discharge data, cannot be anonymized through deidentification. The reason that removing identifiers does not assure sufficient anonymity of respondents is
that, today, a data snooper can get inexpensive access to databases
with names attached to records. Marketing and credit information
databases and voter registration lists are exemplars. Having this external information, the data snooper can employ sophisticated, but readily
available, record linkage techniques. The resultant attempts to link an
identified record from the public database to a deidentified record are
often successful (Winkler, 1998). With such a linkage, the record would
be *reidentified*.

New Areas of Concern

Technological developments continue to raise new issues that must
be addressed in the ethical direction of research involving human subjects. Of burgeoning importance in recent years are developments in
information technology, especially the Internet, and in biotechnology,
especially human genetics research.

The Internet

A good discussion of some of the issues involved in providing remote access to data through the web is provided by Blakemore (2001).
These include security assurances against hacker attack and fears of

record linkage. A prominent example of web access to data is American FactFinder, maintained by the U.S. Census Bureau (http://factfinder. census.gov). American FactFinder provides access to population, housing, economic, and geographic data. The site gives a good description of the elaborate procedures followed to ensure confidentiality through statistical disclosure limitation (see also American Association for the Advance of Science, 1999).

Genetic Research

The American Society of Human Genetics published the following statement on this issue:

> Studies that maintain identified or identifiable specimens must maintain subjects' confidentiality. Information from these samples should not be provided to anyone other than the subjects and persons designated by the subjects in writing. To ensure maximum privacy, it is strongly recommended that investigators apply to the Department of Health and Human Services for a Certificate of Confidentiality. ... Investigators should indicate to the subject that they cannot guarantee absolute confidentiality.

A statement by the Health Research Council of New Zealand (1998) is more specific:

> Researchers must ensure the confidentiality and privacy of stored genetic information, genetic material or results of the research which relate to identified or identifiable participants. In particular, the research protocol must specify whether genetic information or genetic material and any information derived from studying the genetic material, will be stored in identified, deidentified or anonymous form. Researchers should consider carefully the consequences of storing information and material in anonymous form for the proposed research, future research and communication of research results to participants. Researchers should disclose where storage is to be and to whom their tissues will be accessible. *Tissue or DNA should only be sent abroad if this is acceptable to the consenting individual.*

TENSION BETWEEN DISCLOSURE RISK AND DATA UTILITY

Data Quality Audit

The process of assuring confidentiality through statistical disclosure limitation while maintaining data utility has the following components:

- a data quality audit that, beginning with the original, collected data, assesses disclosure risk and data utility;

- a determination of adequacy of confidentiality protection;

- if confidentiality protection is inadequate, the implementation of a restricted access or restricted data procedure; and

- a return to the data quality audit.

A quality audit of collected data evaluates the utility of the data and assesses disclosure risk. Typically, with good research design and implementation, the data utility is high. But, also, the risk of disclosure through the release of the original, collected data is too high, even when the data collected have been deidentified, i.e., apparent identifiers (name, e-mail address, phone number, etc.) have been removed. Reidentification techniques have become too sophisticated to assure confidentiality protection (Winkler, 1998). A confidentiality audit will include identification of (1) sensitive objects and (2) characteristics of the data that make it susceptible to attack.

R-U Confidentiality Map

A measure of statistical disclosure risk, R, is a numerical assessment of the risk of unintended disclosures following dissemination of the data. A measure of data utility, U, is a numerical assessment of the usefulness of the released data for legitimate purposes. Illustrative results using particular specifications for R and U have been developed. The *R-U confidentiality map* was initially presented by Duncan and Fienberg (1999) and further explored for categorical data by Duncan et al. (2001). As it is more fully developed by Duncan, Keller-McNulty, and Stokes (2002), the R-U confidentiality map provides a quantified link between R and U directly through the parameters of a disclosure limitation procedure. With an explicit representation of how the parameters of the disclosure limitation procedure affect R and U, the tradeoff between disclosure risk and data utility is apparent. With the R-U confidentiality map, data-holding groups have a workable new tool to frame decision making about data dissemination under confidentiality constraints.

Restricted Access Procedures

Restricted access procedures are administrative controls on who can access data and under what conditions. These controls may include use of sworn agent status, licensing, and secure research sites. Each of these restricted access procedures requires examination of its structure and careful monitoring to ensure that it provides both confidentiality protection and appropriate access to data. Licensing systems, for example, require periodic inspections and a tracking database to monitor restricted-use data files (Seastrom, 2001). Even in secure research sites, only restricted data may be made available, say with de-identified data files. Secure sites require a trained staff who can impart a "culture of confidentiality" (Dunne, 2001).

Restricted Data Procedures: Disclosure Limitation Methods

Restricted data procedures are methods for disclosure limitation that require a disseminated data product to be some transformation of the original data. A variety of disclosure limitation methods have been proposed by researchers on confidentiality protection. Generally, these methods are tailored either to tabular data or to microdata. These procedures are widely applied by government statistical agencies since they face confidentiality issues directly in producing data products for their users. The most commonly used methods for tabular data are cell suppression based on minimum cell count or dominance rules; recoding variables; rounding; and geographic or minimum population thresholds. The most commonly used methods for microdata are microaggregation, deletion of data items, deletion of sensitive records, recoding data into broad categories, top and bottom coding, sampling, and geographic or minimum population thresholds (see Felsö, Theeuwes, and Wagner, 2001).

Direct transformations of data for confidentiality purposes are called *disclosure limiting masks* (Jabine, 1993a, 1993b). With masked data sets, there is a specific functional relationship, possibly as a function of multiple records and possibly as a stochastic function, between masked values and the original data. Because of this relationship, the possibilities of both identity and attribute disclosures continue to exist, even though the risk of disclosure may be substantially reduced. The idea is to provide a response that, while useful for statistical analysis purposes, has sufficiently low disclosure risk. As a general classification, disclosure-limiting masks can be categorized as suppressions, recodings, or samplings.

Whether for microdata or tabular data, many of these transformations can be represented as matrix masks (Duncan and Pearson, 1991), $M = AXB + C$, where X is a data matrix, say $n \times p$. In general, the defining matrices A, B, and C can depend on the values of X and be stochastic. The matrix A (since it operates on the rows of X) is a record-transforming mask, the matrix B (since it operates on the columns of X) is a variable-transforming mask, and the matrix C is a displacing mask (noise addition).

Methods for Tabular Data

A variety of disclosure limitation methods for tabular data are identified or developed and then analyzed by Duncan et al. (2001). The discussion below tells about some of the more important of these methods.

Suppression

A suppression is a refusal to provide a data instance. For microdata, this can involve the deletion of all values of some particularly sensitive variable. In principle, certain record values could also be suppressed, but this is usually handled through recoding. For tabular data, the values of table cells that pose confidentiality problems are suppressed. These are the primary suppressions. Often, a cell is considered unsafe for publication according to the (n, p) dominance rule, i.e., if a few (n), say three, contributing entities represent a percentage p, say 70 percent, or more of the total. Additionally, enough other cells are suppressed so that the values of the primary suppressions cannot be inferred from released table margins. These additional cells are called secondary suppressions. Even tables of realistic dimensionality with only a few primary suppressions present a multitude of possible configurations for the secondary cell suppressions. This raises computational difficulties that can be formulated as combinatorial optimization problems. Typical techniques that are used include mathematical programming (especially integer programming) and graph theory (Chowdhury et al., 1999).

Recoding

A disclosure-limiting mask for recoding creates a set of data for which some or all of the attribute values have been altered. Recoding can be applied to microdata or to tabular data. Some common methods of recoding for tabular data are global recoding and rounding. A new method of recoding is Markov perturbation.

- Under *global recoding*, categories are combined. This represents a coarsening of the data through combining rows or combining columns of the table.

- Under *rounding*, every cell entry is rounded to some base b. The controlled rounding problem is to find some perturbation of the original entries that will satisfy (marginal, typically) constraints and that is "close" to the original entries (Cox, 1987). Multidimensional tables present special difficulties. Methods for dealing with them are given by Kelley, Golden, and Assad (1990).

- *Markov perturbation* (Duncan and Fienberg, 1999) makes use of stochastic perturbation through entity moves according to a Markov chain. Because of the cross-classified constraints imposed by the fixing of marginal totals, moves must be coupled. This coupling is consistent with a Gröbner basis structure (Fienberg, Makov, and Steele 1998). In a graphical representation, it is consistent with data flows corresponding to an alternating cycle, as discussed by Cox (1987).

Disclosure-Limitation Methods for Microdata

Examples of recoding as applied to microdata include data swapping; adding noise; and global recoding and local suppression. In data swapping (Dalenius and Reiss, 1982; Reiss, 1980; Spruill, 1983), some fields of a record are swapped with the corresponding fields in another record. Concerns have been raised that while data swapping lowers disclosure risk, it may excessively distort the statistical structure of the original data (Adam and Wortmann, 1989). A combination of data swapping with additive noise has been suggested by Fuller (1993). Masking through the introduction of additive or multiplicative noise has been investigated (e.g., Fuller, 1993). A disclosure limitation method for microdata that is used in the μ-Argus software is a combination of global recoding and local suppression. Global recoding combines several categories of a variable to form less specific categories. Topcoding is a specific example of global recoding. Local suppression suppresses certain values of individual variables (Willenborg and de Waal, 1996). The aim is to reduce the set of records where only a few agree on particular combinations of key values. Both methods make the data less specific and so result in some information loss to legitimate researchers.

Sampling

Sampling, as a disclosure-limiting mask, creates an appropriate statistical sample of the original data. Alternatively, if the original data is itself a sample, the data may be considered self-masked. Just the fact that the data are a sample may not result in disclosure risk sufficiently low to permit data dissemination. In that case, subsampling may be required to obtain a data product with adequately low disclosure risk.

Synthetic, Virtual, or Model-Based Data

The methods described so far have involved perturbations or masking of the original data. These are called *data-conditioned* methods by Duncan and Fienberg (1999). Another approach, while less studied, should be conceptually familiar to statisticians. Consider the original data to be a realization according to some statistical model. Replace the original data with samples (the synthetic data) according to the model. Synthetic data sets consist of records of individual synthetic units rather than records the agency holds for actual units.

Rubin (1993) suggested synthetic data construction through a multiple imputation method. The effect of imputation of an entire microdata set on data utility is an open research question. Rubin (1993) asserts that the risk of identity disclosure can be eliminated through the dissemination of synthetic data and proposes the release of synthetic microdata sets for public use. His reasoning is that the synthetic data carries no direct functional link between the original data and the disseminated data. So while there can be substantial identity disclosure risk with (inadequately) masked data, identity disclosure is, in a strict sense, impossible with the release of synthetic data. However, the release of synthetic data may still involve risk of attribute disclosure (Fienberg, Makov, and Steele 1998).

Rubin (1993) cogently argues that the release of synthetic data has advantages over other data dissemination strategies, because

- masked data can require special software for its proper analysis for each combination of analysis, masking method, and database type (Fuller, 1993);

- release of aggregates, e.g., summary statistics or tables, is inadequate due of the difficulty in contemplating at the data release stage what analysts might like to do with the data; and

- mechanisms for the release of microdata under restricted access conditions, e.g., user-specific administrative controls, can never fully satisfy the demands for publicly available microdata.

The methodology for the release of synthetic data is simple in concept, but complex in implementation. Conceptually, the data-holding research group would use the original data to determine a model to generate the synthetic data. But the purpose of this model is not the usual prediction, control, or scientific understanding that argues for parsimony through Occam's Razor. Instead, its purpose is to generate synthetic data useful to a wide range of users. The agency must recognize uncertainty in both model form and the values of model parameters. This argues for the relevance of hierarchical and mixture models to generate the synthetic data.

CONCLUSIONS

IRBs must examine protocols for human subjects research carefully to ensure that both confidentiality protection is afforded and that appropriate data access is afforded. Promising procedures are available based on restricted access, through means such as licensing and secure research sites, and restricted data, through statistical disclosure limitation.

REFERENCES AND BIBLIOGRAPHY

Abowd, J.M., and S.D. Woodcock
  2001 Disclosure limitation in longitudinal linked data. Pp. 215-277 in *Confidentiality, Disclosure, and Data Access: Theory and Practical Applications for Statistical Agencies*, P. Doyle, J.I. Lane, J.J.M. Theeuwes, and L.V. Zayatz, eds. Amsterdam: North-Holland/Elsevier.
Adam, N.R., and J.C. Wortmann
  1989 Security-control methods for statistical databases: A comparative study. *ACM Computing Surveys* 21:515-556.
Agarwal, R., and R. Srikant
  2000 Privacy-preserving data mining. *Proceedings of the 2000 ACM SIGMOD on Management of Data*, May 15-18, Dallas, Tex.
American Association for the Advancement of Science
  1999 *Ethical and Legal Aspects of Human Subjects Research on the Internet*. Workshop Report. Available: http://www.aaas.org/spp/dspp/sfrl/projects/intres/report.pdf [4/12/02].
American Medical Informatics Association
  2000 Letter to the U.S. Department of Health and Human Services. Available: http://www.amia.org/resource/policy/nprm_response.html [4/1/03].
Blakemore, M.
  2001 The potential and perils of remote access. Pp. 315-340 in *Confidentiality, Disclosure, and Data Access: Theory and Practical Applications for Statistical Agencies*, P. Doyle, J.I. Lane, J.J.M. Theeuwes, and L.V. Zayatz, eds. Amsterdam: North-Holland/Elsevier.

Chowdhury, S.D., G.T. Duncan, R. Krishnan, S.F. Roehrig, and S. Mukherjee
  1999 Disclosure detection in multivariate categorical databases: Auditing confidentiality protection through two new matrix operators. *Management Science* 45:1710-1723.
Cox, L.H.
  1980 Suppression methodology and statistical disclosure control. *Journal of the American Statistical Association* 75:377-385.
  1987 A constructive procedure for unbiased controlled rounding. *Journal of the American Statistical Association* 82:38-45.
Dalenius, T.
  1986 Finding a needle in a haystack. *Journal of Official Statistics* 2:329-336.
  1988 *Controlling Invasion of Privacy in Surveys.* Department of Development and Research. Statistics Sweden.
Dalenius, T., and S.P. Reiss
  1982 Data-swapping: A technique for disclosure control. *Journal of Statistical Planning and Inference* 6:73-85.
De Waal, A.G., and L.C.R.G. Willenborg
  1996 A view on statistical disclosure for microdata. *Survey Methodology* 22:95-103.
Domingo-Ferrer, J. and V. Torra
  2001 A quantitative comparison of disclosure control methods for microdata. Pp. 111-134 in *Confidentiality, Disclosure, and Data Access: Theory and Practical Applications for Statistical Agencies*, P. Doyle, J.I. Lane, J.J.M. Theeuwes, and L.V. Zayatz, eds. Amsterdam: North-Holland/Elsevier.
Duncan, G.T.
  2001 Confidentiality and statistical disclosure limitation. In N.J. Smelser and P.B. Baltes, eds., *International Encyclopedia of the Social and Behavioral Sciences*. Oxford, England: Elsevier Science.
Duncan, G.T., and S.E. Fienberg
  1999 Obtaining information while preserving privacy: A Markov perturbation method for tabular data. Eurostat. *Statistical Data Protection '98* Lisbon 351-362.
Duncan, G.T., and S. Kaufman
  1996 Who should manage information and privacy conflicts?: Institutional design for third-party mechanisms. *The International Journal of Conflict Management* 7:21-44.
Duncan, G.T., and D. Lambert
  1986 Disclosure-limited data dissemination (with discussion). *Journal of the American Statistical Association* 81:10-28.
  1989 The risk of disclosure of microdata. *Journal of Business and Economic Statistics* 7:207-217.
Duncan, G.T., and S. Mukherjee
  2000 Optimal disclosure limitation strategy in statistical databases: Deterring tracker attacks through additive noise. *Journal of the American Statistical Association* 95:720-729.
Duncan, G.T., and R. Pearson
  1991 Enhancing access to microdata while protecting confidentiality: Prospects for the future (with discussion). *Statistical Science* 6:219-239.
Duncan, G.T., S.E. Fienberg, R. Krishnan, R. Padman, and S.F. Roehrig
  2001 Disclosure limitation methods and information loss for tabular data. Pp. 135-166 in *Confidentiality, Disclosure, and Data Access: Theory and Practical Applications for Statistical Agencies*, P. Doyle, J.I. Lane, J.J.M. Theeuwes, and L.V. Zayatz, eds. Amsterdam: North-Holland/Elsevier.

Duncan, G.T., S. Keller-Mcnulty, and S.L. Stokes
    2002 Disclosure risk vs. data utility: The R-U confidentiality map. Technical Re-
        ports: Statistical Sciences Group, Los Alamos National Laboratory and Heinz
        School of Public Policy and Management, Carnegie Mellon University.
Dunne, T.
    2001 Issues in the establishment and management of secure research sites. Pp.
        297-314 in *Confidentiality, Disclosure, and Data Access: Theory and Practical
        Applications for Statistical Agencies*, P. Doyle, J.I. Lane, J.J.M. Theeuwes, and
        L.V. Zayatz, eds. Amsterdam: North-Holland/Elsevier.
Elliot, M.
    2001 Disclosure risk assessment. Pp. 135-166 in *Confidentiality, Disclosure, and
        Data Access: Theory and Practical Applications for Statistical Agencies*, P. Doyle,
        J.I. Lane, J.J.M. Theeuwes, and L.V. Zayatz, eds. Amsterdam: North-Holland/
        Elsevier.
Elliot, M., and A. Dale
    1999 Scenarios of attack: The data intruder's perspective on statistical disclosure
        risk. *Netherlands Official Statistics* 14:6-10.
Elliot, M., C. Skinner, and A. Dale
    1998 Special uniques, random uniques and sticky populations: Some counterin-
        tuitive effects of geographical detail on disclosure risk. *Research in Official
        Statistics* 1:53-68.
Eurostat
    1996 *Manual on Disclosure Control Methods.* Luxembourg: Office for Publications
        of the European Communities.
Federal Committee on Statistical Methodology
    1994 Statistical Policy Working Paper 22: Report on Statistical Disclosure Limita-
        tion Methodology. Washington, DC: U.S. Office of Management and Budget.
Felsö, F., J. Theeuwes, and G. Wagner
    2001 Disclosure limitation methods in use: Results of a survey. Pp. 17-42 in *Confi-
        dentiality, Disclosure, and Data Access: Theory and Practical Applications for
        Statistical Agencies*, P. Doyle, J.I. Lane, J.J.M. Theeuwes, and L.V. Zayatz, eds.
        Amsterdam: North-Holland/Elsevier.
Fienberg, S.E.
    1994 Conflicts between the needs for access to statistical information and demands
        for confidentiality. *Journal of Official Statistics* 10:115-132.
Fienberg, S.E., U.E. Makov, and R.J. Steele
    1998 Disclosure limitation using perturbation and related methods for categorical
        data. *Journal of Official Statistics* 14:347-360.
Fuller, W.A.
    1993 Masking procedures for microdata disclosure limitation. *Journal of Official
        Statistics* 9:383-406.
Greenberg, B.
    1990 Disclosure avoidance research at the Census Bureau. Pp. 144-166 in *Pro-
        ceedings of the U.S. Census Bureau Annual Research Conference*, Washington,
        DC.
Greenberg, B. and L. Zayatz
    1992 Strategies for measuring risk in public use microdata files. *Statistica Neer-
        landica* 46:33-48.
Health Research Council of New Zealand
    1998 Statement. Available: http://www.hrc.govt.nz/genethic.htm.
Jabine, T.B.
    1993a Procedures for restricted data access. *Journal of Official Statistics* 9:537-589.
    1993b Statistical disclosure limitation practices of United States statistical agencies.
        Journal of Official Statistics 9:427-454.

Kelley, J., B. Golden, and A. Assad
    1990 Controlled rounding of tabular data. *Operations Research* 38:760-772.
Kim, J.J.
    1986 A method for limiting disclosure in microdata based on random noise and transformation. Pp. 370-374 in *Proceedings of the Survey Research Methods Section*, American Statistical Association.
Kim, J.J., and W. Winkler
    1995 Masking microdata files. In *Proceedings of the Section on Survey Research Methods*, American Statistical Association.
Kooiman, P., J. Nobel, and L. Willenborg
    1999 Statistical data protection at Statistics Netherlands. *Netherlands Official Statistics* 14:21-25.
Lambert, D.
    1993 Measures of disclosure risk and harm. *Journal of Official Statistics* 9:313-331.
Little, R.J.A.
    1993 Statistical analysis of masked data. *Journal of Official Statistics* 9:407-426.
Marsh, C., C. Skinner, S. Arber, B. Penhale, S. Openshaw, J. Hobcraft, D. Lievesley, and N. Walford
    1991 The case for samples of anonymized records from the 1991 census. *Journal of the Royal Statistical Society, Series A* 154:305-340.
Marsh, C., A. Dale, and C.J. Skinner
    1994 Safe data versus safe settings: Access to microdata from the British Census. *International Statistical Review* 62:35-53.
Mokken, R.J., P. Kooiman, J. Pannekoek, and L.C.R.J. Willenborg
    1992 Disclosure risks for microdata. *Statistica Neerlandica* 46:49-67.
Mood, A.M., F.A. Graybill, and D.C. Boes
    1963 *Introduction to the Theory of Statistics*. New York: McGraw-Hill.
Moore, R.A.
    1996 Controlled data-swapping techniques for masking public use microdata sets. *Statistical Research Division Report Series, RR 96-04*. Washington, DC: U.S. Bureau of the Census.
National Research Council
    1993 *Private Lives and Public Policies: Confidentiality and Accessibility of Government Statistics*. Panel on Confidentiality and Data Access, G.T. Duncan, T.B. Jabine, and V.A. de Wolf, eds. Committee on National Statistics and Social Science Research Council. Washington, DC: National Academy Press.
    2000 *Improving Access to and Confidentiality of Research Data: Report of a Workshop* Committee on National Statistics, C. Mackie and N. Bradburn, eds. Washington, DC: National Academy Press.
Paass, G.
    1988 Disclosure risk and disclosure avoidance for microdata. *Journal of Business and Economic Statistics* 6:487-500.
Rubin, D.B.
    1993 Satisfying confidentiality constraints through the use of synthetic multiply-imputed microdata. *Journal of Official Statistics* 9:461-468.
Seastrom, M.M.
    2001 Licensing. Pp. 279-296 in *Confidentiality, Disclosure, and Data Access: Theory and Practical Applications for Statistical Agencies*, P. Doyle, J.I. Lane, J.J.M. Theeuwes, and L.V. Zayatz, eds. Amsterdam: North-Holland/ Elsevier.
Skinner, C.J.
    1990 *Statistical Disclosure Issues for Census Microdata*. Paper presented at International Symposium on Statistical Disclosure Avoidance, Voorburg, The Netherlands, December 13.

Spruill, N.L.
  1983 The confidentiality and analytic usefulness of masked business microdata. Pp. 602-607 in *Proceedings of the Section on Survey Research Methods*, American Statistical Association.
Sweeney, L.
  2001 Information explosion. Pp. 43-74 in *Confidentiality, Disclosure, and Data Access: Theory and Practical Applications for Statistical Agencies*, P. Doyle, J.I. Lane, J.J.M. Theeuwes, and L.V. Zayatz, eds. Amsterdam: North-Holland/ Elsevier.
Willenborg, L., and T. de Waal
  1996 *Statistical Disclosure Control in Practice*. Lecture Notes in Statistics #111. New York: Springer.
Winkler, W.E.
  1998 Re-identification methods for evaluating the confidentiality of analytically valid microdata. *Research in Official Statistics* 1:87-104.
Zayatz, L.V., P. Massell, and P. Steel
  1999 Disclosure limitation practices and research at the U.S. Census Bureau. *Netherlands Official Statistics* 14:26-29.

Biographical Sketches of Panel Members and Staff

Cora B. Marrett, *Chair*, joined the University of Wisconsin system as Senior Vice President for Academic Affairs in 2001, following 4 years as Vice Chancellor for Academic Affairs and Provost at the University of Massachusetts-Amherst. She has held faculty appointments at the University of Wisconsin-Madison, Western Michigan University, and the University of North Carolina at Chapel Hill. Previously, she served as the first Assistant Director for the Social, Behavioral, and Economic Sciences at the National Science Foundation and as the director of two programs for the United Negro College Fund under a grant from the Andrew Mellon Foundation. She holds a B.A. degree from Virginia Union University and M.A. and Ph.D. degrees from the University of Wisconsin-Madison, all in sociology. She served as a member of the Division of Social and Behavioral Sciences and Education at the National Academies and is currently on the boards for the Argonne National Laboratory, the Russell Sage Foundation, and the Social Science Research Council. She is a fellow of the American Association for the Advancement of Science, the American Academy of Arts and Sciences, and Sigma Xi, the Science Research Society.

Daniel R. Ilgen, *Vice Chair*, is the John A. Hannah professor of psychology and management at Michigan State University and has served on the university's institutional review board. Previously, he was a member of the Industrial and Organizational Psychology Department faculty at Purdue University. He is editor of *Organizational Behavior and Human Decision Processes*. His areas of specialization are in work motivation and small group/team behavior, particularly team decision making, and he is currently a member of the National Research Council's Committee on Human Factors. He received a Ph.D. degree in psychology from the University of Illinois, Urbana-Champaign.

Tora Kay Bikson is a senior scientist in the Behavioral Sciences Department of RAND, and she chairs RAND's institutional review board. Her research has investigated properties of advancing information technologies in varied user contexts, addressing such issues as what factors affect the successful incorporation of innovative tools in ongoing activities and how new work media influence group structures and interaction processes. Previously, she has taught courses on ethical issues in human subjects research in the RAND Graduate School and in the Honors College at the University of California at Los Angeles. She has also served on committees concerned with information technology and with data privacy at the National Academies. She received a Ph.D. degree in philosophy from the University of Missouri.

Jamie L. Casey is a research assistant for the Committee on National Statistics (CNSTAT). She has worked on CNSTAT projects studying the Special Supplemental Nutrition Program for Women, Infants, and Children (WIC) eligibility, the State Children's Health Insurance Program, and the 2000 Census. Previously, she worked for the National Center for Health Statistics. She received a B.A. degree in psychology from Goucher College.

Constance F. Citro, *Study Director*, is a senior program officer for the Committee on National Statistics. She is a former vice president and deputy director of Mathematica Policy Research, Inc., and was an American Statistical Association/National Science Foundation research fellow at the U.S. Census Bureau. For the committee, she has served as study director for numerous projects, including the Panel to Evaluate the 2000 Census, the Panel on Poverty and Family Assistance, the Panel to Evaluate the Survey of Income and Program Participation, and the Panel to Evaluate Microsimulation Models for Social Welfare Programs. Her research focuses on the quality and accessibility of large, complex microdata files and analysis related to income and poverty measurement. She is a fellow of the American Statistical Association. She received a B.A. degree from the University of Rochester and M.A. and Ph.D. degrees in political science from Yale University.

Robert M. Groves is director of the Survey Research Center, a professor of sociology, and senior research scientist at the Institute for Social Research, all at the University of Michigan. Previously he was the associate director and then director of the Joint Program in Survey Methodology, based at the University of Maryland, a consortium of the University of Maryland, the University of Michigan, and Westat, Inc., sponsored by the federal statistical system. He also served as associate director of the U.S. Census Bureau in 1990-1992. He has investigated the effects of alternative telephone sample designs on precision, the effect of data collection mode on the quality of survey reports, causes and remedies for nonresponse errors in surveys, estimation and explanation of interviewer variance in survey responses, and other topics in survey methods. His current research interests focus on theory-building in survey participation and models of nonresponse reduction and adjustment. He is a member of the Committee on National Statistics and a fellow of the American Statistical Association. He has an A.B. degree from Dartmouth College and a Ph.D. degree from the University of Michigan.

Robert M. Hauser is Vilas research professor of sociology at the University of Wisconsin-Madison, where he directs the Center for Demography of Health and Aging and the Wisconsin Longitudinal Study. During 2001-2002 he was a visiting scholar at the Russell Sage Foundation. He is a member of the National Academy of Sciences and is a fellow of the National Academy of Education, the American Association for the Advancement of Science, the American Statistical Association, the Center for Advanced Study in the Behavioral Sciences, and the American Academy of Arts and Sciences. He has served on the National Research Council's Committee on National Statistics, Commission on Behavioral and Social Sciences and Education, and Board on Testing and Assessment, and he chaired the National Research Council's Committee on the Appropriate Use of High Stakes Tests. His current research interests include trends in educational progression and social mobility in the United States among racial and ethnic groups, the uses of educational assessment as a policy tool, the effects of families on social and economic inequality, and changes in socioeconomic standing, health, and well-being across the life course. He received a B.A. degree from the University of Chicago and a Ph.D. degree from the University of Michigan.

V. Joseph Hotz is a professor and chair of the Department of Economics at the University of California at Los Angeles. He was a national research associate of the Northwestern University/University of Chicago Joint Center for Poverty Research and chaired the center's Advisory Panel for Research Uses of Administrative Data. He is a research associate of the National Bureau of Economic Research, a member of the board of overseers of the Panel Study of Income Dynamics, and chair of the oversight board of the California Census Research Data Center. His research focuses on the economics of the family, applied econometrics, and the evaluation of social programs. He received his Ph.D. degree in economics from the University of Wisconsin-Madison.

Tanya M. Lee is a project assistant for the Committee on National Statistics. Before joining CNSTAT, she worked at the National Academies' Institute of Medicine for the Committee on Strategies for Small Number Participants Clinical Research Trials and the Committee on Creating a Vision for Space Medicine during Travel Beyond Earth Orbit. She is pursuing a degree in the field of psychology.

Patricia Marshall is associate professor of bioethics in the Center for Biomedical Ethics at Case Western Reserve University. Previously, she was an associate professor in the Department of Medicine and the Neiswanger Institute of Biomedical Ethics and Health Policy at Loyola University of Chicago. She has served as a consultant to the President's National Bioethics Advisory Commission on a project examining ethical issues in international health research and as a consultant to the World Health Organization's Council for International Organization of Medical Societies on their revision of ethical guidelines for international research. Her research interests and publications focus on multiculturalism and the application of bioethics practices, research ethics and informed consent, and HIV prevention among injection drug users. She has a B.A. degree in behavioral science and M.A. and Ph.D. degrees in anthropology from the University of Kentucky.

Anna C. Mastroianni is assistant professor at the School of Law and the Institute for Public Health Genetics at the University of Washington. She also holds appointments in the Department of Health Services in the university's School of Public Health and Community Medicine and in the Department of Medical History and Ethics in the School of Medicine. She is a Greenwall Foundation faculty scholar in bioethics and a center associate at the University of Minnesota's Center for Bioethics. Her research and teaching is in the area of health law and bioethics, with specific interests in legal, ethical, and policy issues related to human subjects research, the use of genetic technologies, women's health, reproductive rights, the use of assisted reproductive technologies, and the responsible conduct of research. She has held a number of legal and federal policy positions, including associate director of the White House Advisory Committee on Human Radiation Experiments. She is a fellow of the American Association for the Advancement of Science. She holds a J.D. degree from the University of

Pennsylvania School of Law, a B.S. degree in economics from the university's Wharton School, and a B.A. degree in Spanish and Portuguese from the university's College of Arts and Sciences, as well as an M.P.H. degree from the University of Washington School of Public Health and Community Medicine.

John J. (Jack) McArdle is professor of psychology at the University of Virginia. He is also director of the Jefferson Psychometric Laboratory, a visiting fellow at the Institute of Human Development at University of California at Berkeley, an adjunct faculty member at the Department of Psychiatry at the University of Hawaii, and the lead data analyst for research studies on college student-athletes at the National Collegiate Athletic Association. He received a B.A. degree in psychology and mathematics at Franklin and Marshall College and M.A. and Ph.D. degrees in psychology and computer sciences at Hofstra University. His research focuses on age-sensitive methods for psychological and educational measurement and longitudinal data analysis. He has published work in factor analysis, growth curve analysis, and dynamic modeling of adult cognitive abilities.

Eleanor Singer is associate director and senior research scientist at the Survey Research Center of the Institute for Social Research and an adjunct professor of sociology at the University of Michigan. Previously, she was a senior research scholar at the Center for Social Sciences at Columbia University. She has served as president of the American Association for Public Opinion Research and as chair of its ethics committee, as well as editor of *Public Opinion Quarterly*. Her research has largely focused on methodological issues in surveys, among them the effect of confidentiality concerns on survey participation. She received a B.A. degree from Queens College and a Ph.D. degree from Columbia University.

William A. Yost, *Liaison, Board on Behavioral, Cognitive, and Social Sciences*, is associate vice president for research and dean of the graduate schools and professor of hearing sciences at Loyola University of Chicago. He was previously the director of the Parmly Hearing Institute and director of the interdisciplinary neuroscience minor program at Loyola. He also is an adjunct professor of psychology, adjunct professor of otolaryngology, and a member of the Parmly Hearing Institute. He received a B.S. degree in psychology from the Colorado College and a Ph.D. degree in Experimental Psychology from Indiana

University, and he received an honorary degree of doctor of science from the Colorado College. He has served on the faculty at the University of Florida and held visiting appointments at Northwestern University and the Colorado College. His specialty within the area of hearing sciences is auditory perception and psychoacoustics. He is a fellow of the Acoustical Society of America, the American Speech, Hearing, and Language Association, the American Psychological Society, and the American Association for the Advancement of Science.